SOLID-STATE RADAR TRANSMITTERS

The Artech House Radar Library

Radar System Analysis by David K. Barton

Electronic Intelligence: The Analysis of Radar Signals by Richard G. Wiley

Electronic Intelligence: The Interception of Radar Signals by Richard G. Wiley

Principles of Secure Communication Systems by Don J. Torrieri

Shipboard Antennas by Preston E. Law, Jr.

Radar Propagation at Low Altitudes by M.L. Meeks

Interference Suppression Techniques for Microwave Antennas and Transmitters by Ernest R. Freeman

Radar Cross Section by Eugene F. Knott, John F. Shaeffer, and Michael T. Tuley

Radar Anti-Jamming Techniques by M.V. Maksimov et al.

Synthetic Array and Imaging Radars by S.A. Hovanessian

Radar Detection and Tracking Systems by S.A. Hovanessian

Radar System Design and Analysis by S.A. Hovanessian

Radar Calculations Using the TI-59 Programmable Calculator by William A. Skillman

Radar Calculations Using Personal Computers by William A. Skillman

Techniques of Radar Reflectivity Measurement, N.C. Currie, ed.

Monopulse Principles and Techniques by Samuel M. Sherman

Receiving Systems Design by Stephen J. Erst

Signal Theory and Random Processes by Harry Urkowitz

Radar Reflectivity of Land and Sea by M.W. Long

High Resolution Radar Imaging by Dean L. Mensa

Introduction to Monopulse by Donald R. Rhodes

Probability and Information Theory, with Applications to Radar by P.M. Woodward

Radar Detection by J.V. DiFranco and W.L. Rubin

RF Radiometer Handbook by C.W. McLeish and G. Evans

Synthetic Aperture Radar, John J. Kovaly, ed.

Infrared-To-Millimeter Wavelength Detectors, Frank R. Arams, ed.

Significant Phased Array Papers, R.C. Hansen, ed.

Phased Array Antennas, A. Oliner and G. Knittel, eds.

Handbook of Radar Measurement by David K. Barton and Harold R. Ward

Statistical Theory of Extended Radar Targets by R.V. Ostrovityanov and F.A. Basalov

Antennas by Lamont V. Blake

Radars, 7 vol., David K. Barton, ed.

SOLID-STATE RADAR TRANSMITTERS

Edward D. Ostroff
Michael Borkowski
Harry Thomas
James Curtis

Foreword

Solid-state devices have been used in radar ever since the 1940s, when the crystal detector served as the input mixer for most microwave receiver systems. Application of the transistor brought solid-state into the amplifier stages of the receiver and signal processor, during the 1960s, while digital technology has caused almost universal replacement of vacuum tubes with integrated circuits in the signal and data processing subsystems since 1970. Solid-state microwave sources have since found their way into the receiving system as local oscillators, while power supplies were converted to solid-state operation during the same period. In most of today's radars, however, the transmitter and the display subsystems continue to use vacuum tubes.

Now it is time to consider the high-power, high-frequency transistor as the source of transmitter power, replacing the triode, the magnetron, the klystron, and the traveling-wave tube. In this book, the technology and design procedure for introducing transistors into radar transmitters are discussed in a systematic way for the first time, by experts who have actually participated in the process. Starting from the level of the junction and the packaged device, the characteristics of solid-state radar transmitters are developed through the power amplifier circuit, the combination of amplifiers into modules, and the final summing of output power in space or in an output combiner. Considerations of power supplies, regulation, reliability, and cost are also discussed. Finally, examples of successful radar system designs using solid-state transmitters are described.

As with most areas of new technology, there has been a lag between the development of devices and circuits using solid-state to produce RF power, and the development of radar systems exploiting these circuits. A major field in which the exploitation has proceeded rapidly has been the modular array, in which each radiating element has its own transit-receive module. For the lower radar frequencies (e.g., 425 MHz), numerous practical designs have been completed, and major systems have been built using these designs. In this frequency band, the solid-state transmitter has also been applied successfully to a conventional radar using a rotating reflector. As radar frequency is increased, so is the difficulty in achieving high powers and high efficiencies needed for long-range operation. At L band, the current designs are limited to those in which very long pulses can be used to achieve the detection of targets at long range. This problem becomes more severe at S band and higher frequencies. Yet, the trend toward improved devices, with higher efficiency and power output, forces the system designer to consider the solid-state transmitter as an alternative to tubes at all radar frequencies for future systems.

Those working in the solid-state device field can use this book as a guide to shaping new designs to meet radar requirements. The radar engineer can consider it a guide to shaping his requirements to the capabilities of the device. What both must keep in mind is that successful radar systems are rare, considering the number of developments which are started and carried through major contract efforts. The choice of waveforms is crucial to the success of the final system, and must generally be based on resolution requirements imposed by the environment, particularly radar clutter. The transmitter must support these waveforms, without introducing more clutter and spurious outputs which destroy the ability of the radar to see small targets. This may require that the solid-state device be designed to produce reasonable average powers with lower duty factors than are now considered convenient: e.g., two to ten percent. To the extent that this can be done, solid-state transmitters will have broad application as replacements for existing tube transmitters and in new radars used for navigation, air search, and other common applications. If their application is limited to radars which can use waveforms of 20 to 30 percent duty factor, solid-state transmitters will not compete successfully in a large number of systems which require that the receiver be sensitive during most of the operating cycle. This is the challenge for designers of solid-state radar transmitters, and the design approaches presented in this book should be helpful in meeting it at higher and higher microwave frequencies over the next decades.

David K. Barton

CONTENTS

Foreword by David K. Barton   *v*

1 INTRODUCTION   1

2 RF POWER TRANSISTORS   5
    2.1 History   5
    2.2 Transistor Dice   7
    2.3 Transistor Packaging   19
    2.4 Peak Power *versus* Average Power   22
    2.5 Reliability Considerations   28
    2.6 Typical Performance of Solid-State Devices   32
    2.7 Specifying Power Transistors for Solid-State Radar Systems   39
    2.8 Trends in Solid-State Devices for Radar Systems   45

3 POWER AMPLIFIER DESIGN   51
    3.1 Introduction   51
    3.2 Impedance Matching   77
    3.3 Design Considerations   107

4 THE CORPORATE STRUCTURE AMPLIFIER   131
    4.1 Properties of the Corporate Structure Amplifier   132
    4.2 Analysis of a Uniform CSA   133
    4.3 Transmitter Efficiency   135
    4.4 Comparison of the PAVE PAWS Transmitter with a CSA   136
    4.5 Partitioning of a CSA into Modules   139
    4.6 Power Output and Rise Time Control   141

5 POWER COMBINER DESIGN   143
    5.1 Power Combiner Types   144
    5.2 Combiner Requirements   145
    5.3 Combiner Selection   145
    5.4 Choice Factors with Regard to Combiner Types   148
    5.5 Power Combiner Designs   149
    5.6 Power Distribution Calculations   154
    5.7 Power Handling   155
    5.8 Coaxial Lines Operating in the TEM Mode   157

5.9   Stripline Operating in the TEM Mode   159
5.10 Microstrip Line Operating in the Dominant Mode   160
5.11 Rectangular Waveguide Operating in the TE_{10} Mode   161
5.12 Ridged Waveguide Operating in the TE_{10} Mode   162
5.13 Component Analysis Using ABCD Matrices   163
5.14 Analysis of a Planar Divider   172

6 POWER AMPLIFIER MODULES   183

7 POWER SUPPLY AND ENERGY STORAGE DESIGN   193
  7.1 Power Supply Options   193
  7.2 Pulse Waveform Amplitude and Phase Variation   195
  7.3 Droop in Pulsed Transmitters   196
  7.4 MTI Improvement Factor Limitations   198

8 RELIABILITY OF SOLID-STATE RADAR TRANSMITTERS   201
  8.1 Reliability Prediction for a CSA   202
  8.2 A Simple Reliability Prediction Equation   206
  8.3 Reliability and Cost Trade-Offs Between Solid-State and Vacuum-Tube Transmitters for Phased Array Radars   206
  8.4 Life Cycle Cost Trade-Off Analysis   212

9 EXAMPLES OF SUCCESSFUL SOLID-STATE RADAR TRANSMITTER DESIGNS   215
  9.1 G.E. TPS-59   217
      The AN/TPS-59, General Electric Company, EHM-12, 413/1000, March 1977   217
      Performance Evaluation by E. J. Gersten and R. A. Joseph, General Electric Company, paper presented at EASCON 76, September 1976   226
      Solid-State Devices for Radar by W. Perkins   230
  9.2 Solid-State Transmit/Receive Module for the PAVE PAWS Phased Array by D. Hoft. Reprinted from *Microwave Journal*, October 1978   246
  9.3 A 250 kW Solid-State AN/SPS-40 Radar Transmitter by K. Lee, C. Corson, and G. Mols. Reprinted from *Microwave Journal*, July 1983   251

APPENDIX: Proof of the CSA Theorems   265

Index   269

Chapter 1

Introduction

The art of designing *solid-state radar transmitters* has evolved rapidly since the mid-1970s and has been accompanied by significant progress in the design of microwave power devices, which include bipolar silicon transistors at frequencies up to S band (approximately 3 GHz) and GaAs FETs at lower power levels in C band and the higher frequency ranges. Rapid progress in the power handling, bandwidth, and reliability of these components will continue. So, any book on the subject is likely to be somewhat out of date before the first printing. This is also true of digital integrated circuit technology. Therefore, the approach we have taken in this book accepts the device's rapidly changing state of the art, and emphasizes the stable and slowly changing technical characteristics common to all high-power transmitters, which are comprised of a large number of relatively low-power devices that have their contributions to the overall output efficiently combined in a power combiner network, or, as in a *phased array,* combined in space.

Unlike the high-power transmitters used for microwave radars, which employed a single exotic vacuum tube such as a *magnetron* or *klystron,* these transmitters employ a large number of active elements. This is sometimes a blessing because it is very improbable that all of the active components will fail at the same time, but it can also be a curse when the cost of many elements adds up to a sum that can often be much larger than the cost of a single large high-power transmitter. The economies of scale derived from large numbers of low-power elements as compared to one or a few high-power elements can only be determined on the basis of a detailed analysis of a particular design. The principles and design approaches presented here should be very useful in such evaluations. High reliability is not ensured by designing the transmitter to consist of a large number of elements rather than a single high-power element. A very conservatively designed high-power transmitter may well be more reliable than a transmitter composed of a large number of overly stressed elements. However, the multistage design will not fail all at once. It will take a very predictable path toward the level at which its performance has degraded beyond a specified acceptable threshold.

The arrangement of the large number of elements needed to produce high output power is of concern in terms of the overall reliability, efficiency, and cost of a transmitter that must deliver a specified peak and average output power. A design approach presented in Chapter 4 assumes that all power amplifier stages are identical and derives certain properties of ensembles of identical amplifier cells, called a *"corporate structure amplifier"* because of analogous use in the description of antenna feed networks in the past, and also because of the similarity between circuit architectures and management hierarchies, which have evolved out of necessity in order to enable a single manager to control the actions of a large number of subordinates with a minimum of intervening levels of sub-management. Similar hierarchical properties apply to communications networks.

Another property of solid-state transmitters, which is uniquely different from the vacuum tube designs, is the dc power supply. Instead of low current and high voltage, the solid-state transmitters operate at low voltage and high current. A *klystron* transmitter that operates at 40 kilovolts at one ampere is likely to be replaced by a solid-state equivalent that operates at 40 volts and 1000 amperes. This raises design issues that encompass the weight and cost of the distribution network needed to feed the large number of devices in an efficient manner. Energy storage in moderate duty cycle designs is another critical issue, particularly where modern radars must provide high phase and amplitude stability in order to "see" targets in a strong clutter background. Power supply voltage fluctuations result in output pulse amplitude and phase fluctuations which can limit the performance of *moving target indicator* (MTI) circuits or other phase coherent anticlutter signal processing.

This book devotes a single chapter to *reliability*, which is a particularly interesting aspect of this type of transmitter. Reliability is influenced by the architecture (the arrangement of replaceable units), as well as the thermal design and operating stress levels of the critical components. As in all solid-state designs, low operating voltages and junction temperatures are important. Transient effects in long pulse designs are also important. Infrared microscope techniques are valuable tools in analyzing operating junction temperature behavior.

Most of the devices of interest operate as class C amplifiers in which drive saturation is not unusual. This results in a power amplifer which has an output power level that is more sensitive to the power supply voltage than it is to the input drive power. It is possible to vary the output level in a controlled way so that the rise and fall time and the amplitude of RF output pulses can be varied to meet desired requirements. This property of the corporate structure amplifier (CSA) transmitter simplifies its output power level control by digital commands using fairly simple switching circuits. The book concludes with a series of examples of successful transmitter designs that have demonstrated the reliability and practicality of this new approach to provide a modular transmitter with high availability, simple maintenance, and, ultimately, low life-cycle cost.

Chapter 2

RF Power Transistors

This chapter outlines the characteristics and requirements of silicon bipolar transistors with regard to their application in solid-state radar systems.

Solid-state radar transmitter design and conversion started in the late 1960s. At that time, silicon bipolar transistors that were designed for general applications, often primarily aimed at the telecommunications market, were used in radar designs. As radar requirements became better understood, transistors were designed such that their characteristics were oriented toward specific radar applications. A brief history of this development follows.

2.1 HISTORY

Microwave power transistors started appearing on the market in the mid-1960s. Early transistors were based on RCA designs, which used aluminum metalization and coaxial packages. Transistors operating above one GHz were usually assembled in the common-base configuration. Below one GHz, stud-mounted transistors with stripline configurations began to be used. Power levels generated by these early devices were moderate compared to those available today. Additionally, the attainable bandwidth of early devices was limited and they were inadequately reliable as well. The following device design changes have contributed to substantial improvements in all of these areas:

1. Improvement in the output power of microwave transistors has come about as the result of a number of things, including changes in die design, package design, and combining techniques. Among these, die design has been the major factor affecting output power performance. Today's power transistors use a cellular die design having sufficient emitter ballast and thermal spread to allow relatively uniform sharing of RF current and heat within the transistor. The uniform sharing of RF current allows larger transistors to be built by placing more active area in parallel inside the package.

Without adequate current sharing and proper thermal layout, additional active area might cause imbalance within the device. This imbalance could result in diminishing returns when active area reaches even relatively modest levels.

2. Improvement in the available bandwidth of microwave transistors has resulted from the incorporation of internal impedance-matching elements inside transistor packages. Today's transistor designs are actually hybrid assemblies consisting of the active transistor chips plus passive internal input and output impedance-matching networks. These networks are designed to prematch the transistor chip impedances, which are low for high power transistors, to levels where the package parasitics no longer limit the amplifier bandwidth. In effect, matching networks are designed to transform chip impedances to system levels, and then portions of those networks are placed inside the package. Matching, therefore, extends into the package to the chip level, and uses package parasitics (principally, the package lead inductance) as part of the total network.

3. The reliability of early microwave transistors was considerably lower than that of most transistors available today. A major factor in the reliability improvement involves transistor metalization and the metal interconnecting system. Early transistors used aluminium die metalization and an aluminium-wire interconnecting system. High current densities, in conjunction with peak junction temperatures, caused metal migration along the metalization stripes. This metal migration caused voids in the metal interconnections and, eventually, transistor failure. In some applications, transistors failed in a matter of weeks. A second cause of failure, especially in pulsed radar systems, was fatigue of the aluminium-wire interconnecting system. Fatigue was caused by repeated expansion and contraction of the bond wires during RF pulsing due to heating as a result of dc loss within the wires. Both metal migration and wire fatigue problems have been substantially eliminated by the conversion to gold-die metalization and an all-gold interconnecting system. Most modern microwave power transistors have excellent reliability histories.

There have been additional changes, which have led to an increased rate of development for solid-state radars as well as radar conversions, such as increases in device performance and reduction of device costs, which are certainly important.

In addition to the developments mentioned above, device performance improvements have resulted from both refinements in processing methods as well as new chip design features. Cost reductions are the result of improvements in processing yields plus a reduction in the direct labor required in device manufacturing. Reduced labor costs are largely due to the implementation of automation in the assembly process. Performance improvements and cost reductions are well established trends in the semiconductor industry.

2.2 TRANSISTOR DICE

The fundamental building block of all microwave power transistors is the semiconductor *chip*. While the chips are all made of doped silicon and consist of interconnected active cells, a number of significantly different geometries and processing techniques are used to produce them. It is beyond the scope of this book to cover all of the different processes and design variations used to manufacture RF power semiconductors. Instead, an overview of device design is included so that the reader may become acquainted with the key elements of the devices.

Silicon bipolar microwave transistors operate on the same principles as other bipolar devices. They are planar and have a vertical construction with the collector on the bottom. Figure 2-1 shows a cross section of the active area of a typical device. This diagram represents a small part of the active area. In fact, it could be the vertical structure of a low-frequency bipolar transistor. The bottom layer, which forms the collector, consists of a highly doped silicon substrate, where a layer of epitaxial silicon, having controlled doping (resistivity), has been grown. Microwave transistors are *NPN* devices because of the higher mobility, and, therefore, higher frequency capability, of electrons. The collector region, therefore, consists of an *N*-doped epitaxial layer grown on an *N+* doped silicon substrate. The properties of the epitaxial layer, in conjunction with subsequent base diffusions, determine the voltage characteristics of the transistor. A *P*-type dopant is either diffused or implanted into the epitaxial region converting the material to a *P* base. Finally, an *N*-type dopant is diffused or implanted into the base to reconvert the material to an *N* emitter.

After all diffusions are completed, ohmic contact is made to the emitter and base regions. The ohmic contact consists of a surface contact plus the metalized interconnecting system. In the case of modern microwave power transistors, the basic metal conductor is gold. Because gold alloys with silicon at low temperatures and would short the transistor junctions, a barrier metal is formed between the silicon surface and the gold conductor. The barrier metal varies among device manufacturers, but it is typically a refractory metal or a metal alloy such as titanium tungsten (TiW).

8

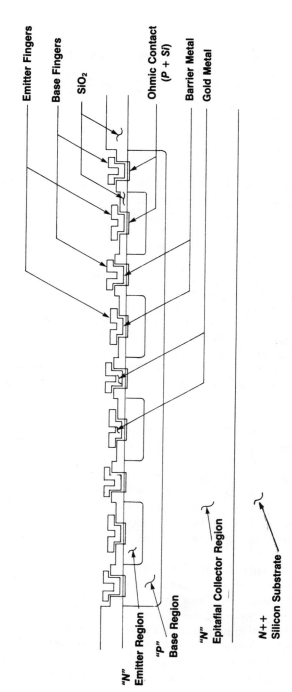

Emitter Fingers

Base Fingers

SiO₂

Ohmic Contact
(P + Si)

Barrier Metal

Gold Metal

"N" Emitter Region

"p" Base Region

"N" Epitafial Collector Region

N++ Silicon Substrate

Figure 2-1 Cross Section of the Active Area of a Multiple-Emitter, Silicon, *NPN*, Microwave Power Transistor

One aspect that distinguishes microwave transistors from their low frequency counterparts is their shallow, narrow base region. This narrow region is primarily responsible for their high $f(t)$. Because the base region is narrow, it has high resistance under the emitter diffusion. This causes most of the current conducted by the device to occur along the periphery of the emitter. In order to maximize current-handling capability, emitter periphery is maximized. This results in a surface geometry with multiple emitter sites, which are eventually contacted and interconnected by metal fingers. The base region is also contacted and interconnected by a metal grid. The grid forms emitter and base fingers, which are interleaved across the surface of the device. Metal must be used to conduct current to the active area in order to reduce the voltage drop of power transistors.

2.2.1 Geometries

The geometry of a device refers primarily to the surface features of the device. Geometries also affect the vertical structure of the chips. There are three principal geometries used in the manufacturing of microwave power transistors: *overlay, interdigitated,* and *matrix.* These terms refer to the structure of the emitter. Each geometry attempts to optimize transistor performance, and there are a number of variations to these basic geometries. In general, performance optimization involves maximizing the amount of emitter periphery within a given base area while minimizing the emitter area within a given amount of base area. The former relates to increasing current-handling capability and the latter relates to minimizing parasitic emitter-to-base capacitance. Other factors are also important, but to a lesser extent. Each of the basic geometries will be briefly described below.

Overlay. The overlay geometry, as shown in Figure 2-2a, consists of a base cell which has two diffusions. The first diffusion is made in order to define the overall base region. A second $P+$ diffusion is made to lower the resistivity of the base region around the multiple overlay emitters. This is done to maximize emitter injection efficiency around the entire emitter periphery. Figure 2-2b shows the addition of diffused emitters into the $P+$ base diffusion. Finally, Figure 2-2c shows the addition of interconnecting metalization. One significant feature of the overlay geometry is that the interconnecting metal fingers are relatively wide, especially the emitter fingers. These wide emitter fingers generally result in lower current densities in the metalization than would be the case for other geometries with equivalent emitter periphery-to-base area (Ep/Ba) ratios. Lower current densities translate into higher reliability with regard

to metal migration failures. This was an important feature of older aluminium metalization systems. For gold metalization, with its significantly
higher resistance to migration, it is much less important.

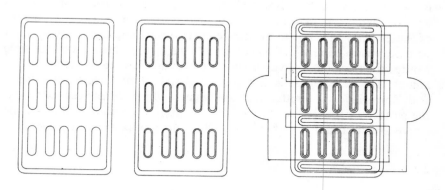

Figure 2-2 Overlay Geometry: (a) Base Diffusion; (b) Added Emitter;
(c) Added Metalization

Interdigitated. The interdigitated geometry uses a simple base diffusion
as shown in Figure 2-3a. Long, narrow emitter sites are diffused into the
base region as shown in Figure 2-3b. Finally, interconnecting metalization
is added, as shown in Figure 2-3c. The interdigitated geometry does not
rely on a low-resistivity diffusion to ensure efficient current injection
around the emitter sites, but rather uses narrowly spaced metal emitter
and base fingers to ensure low resistance current paths. This results in
narrow metal stripes and relatively high current densities. The use of
gold metalization for the conductors and ion-etching techniques to define
metal fingers has resulted in relatively thick metalization, and, therefore,
good resistance to metal migration. Other techniques, such as plated
metal layers, can also be used to reduce current densities. The interdigitated geometry features low emitter area-to-base area ratio, and, therefore, features a high $f(t)$. Generally, it also involves fewer processing
steps than other geometries.

Matrix. The matrix geometry also uses a simple base diffusion as shown
in Figure 2-4a. An emitter having a crosshatched appearance is diffused
into the base region as shown in Figure 2-4b. Finally, contacting fingers
are added as shown in Figure 2-4c. The matrix geometry eliminates the
need for a $P+$ base diffusion, and can achieve moderate emitter periphery-to-base area ratios. The metal fingers are generally wider than those

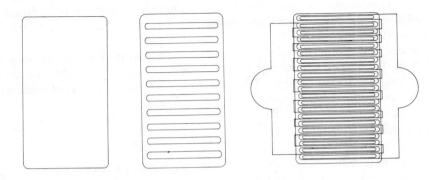

Figure 2-3 Interdigitated Geometry: (a) Base Diffusion; (b) Added Emitter; (c) Added Metalization

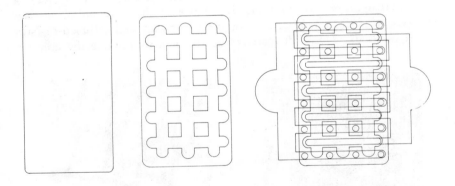

Figure 2-4 Matrix Geometry: (a) Base Diffusion; (b) Added Emitter; (c) Added Metalization

used on interdigitated designs, and, therefore, have acceptable current densities relative to metal migration. This geometry usually has a higher emitter area-to-base area ratio, and, therefore, a somewhat lower $f(t)$. Additionally, the matrix geometry results in metal fingers, which extend over a large number of oxidation steps. This has resulted in metal crack-ing (*micro-cracks*) in some devices using aluminium metalization. Good oxidation step coverage results, however, when a sputtered gold metali-zation system is used.

Basic cell geometries are described in the paragraphs above. In addi-tion to these basic geometries, other elements are designed into modern microwave transistor chips. The two most important consist of *ballast techniques* and *voltage breakdown enhancements*.

2.2.2 Ballast Techniques

Ballast techniques use resistive losses (or feedback) to distribute cur-rent more equally within a transistor chip. Two kinds of ballast techniques are used: emitter ballast and collector ballast. The emitter ballast tech-nique consists of adding a small amount of lumped resistance in series with emitter sites before they are bussed together by interconnecting metalization and bond-wires. Several types of emitter ballast methods are used. The most common method involves an emitter resistor con-nected in series with each emitter finger. This resistor can be either a thin-film resistor or a diffused resistor. Figure 2-5 shows scanning electron microscope (SEM) photographs of both thin-film and diffused-resistor ballast methods. Thin-film resistors are formed by etching the gold (or

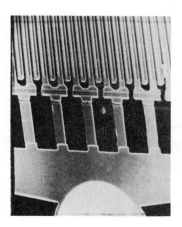 

Figure 2-5 SEM Photographs of Emitter Ballast: (a) Thin-Film Resistors; (b) Diffused Resistors

other conductor metal) off of the emitter metal stripe, leaving the refractory barrier metal, which has high resistivity. Diffused resistors, as the name implies, are formed by a diffusion into the silicon cyrstal, which is contacted by the emitter metalization. The emitter ballast is simply an emitter feedback resistor deposited directly on the transistor chip that forces better current sharing. It also results in lower gain and somewhat higher saturation voltage.

The collector ballast technique is carried out by making the collector epitaxial region thicker than necessary in order to support the designed collector-base depletion. In effect, this results in a resistor in series with the collector regions under each active cell before they are electrically connected by the low-resistivity silicon substrate. The series resistance better distributes current under breakdown conditions. It also limits the current level. Collector ballast, therefore, makes the transistors more rugged under load mismatch, over-voltage, and over-drive conditions. The price paid for this improved performance is lower saturated power and somewhat lower efficiency.

2.2.3 Breakdown Enhancements

The shallow diffusions used to produce microwave power transistors result in high field concentrations around the periphery of the base area. These high field concentrations cause premature breakdown of the collector-base junction, and, therefore, limit the useful operating voltage of the transistors. It is usually desirable to maximize operating voltage because it results in maximum output power. Several techniques are used to enhance breakdowns, all of which result in lower field concentrations at the periphery of the base area where curvature-limited breakdown occurs. One technique involves the diffusion of a relatively deep "moat" around the base cell. This moat reduces the field concentration, and as a result very nearly allows the device to attain the theoretical bulk breakdown limit of the doped crystal.

Each of the geometries described in section 2.2.1 has a number of variations, which are beyond the scope of this chapter. SEM photographs of representative transistor chips using each of the geometries described above are shown in Figure 2-6.

2.2.4 Cellular Construction

Up to this point, microwave transistors have been described in terms of their cell geometries. Early microwave transistors primarily consisted of single cells. Higher power transistors had larger single cells. As power levels (and, therefore, transistor sizes) increased, the concentration of

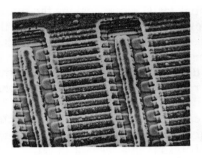
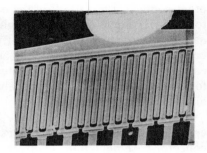

Figure 2-6 SEM Photographs of (a) Overlay, (b) Interdigitated, and (c) Matrix Geometries

thermal energy increased to the point where transistors were thermally limited; that is to say, the active area was capable of supporting high levels of peak power. However, because of the concentration of heat, relatively low levels of average power were attainable.

The thermal design of microwave power transistors is now a major consideration and most transistors are optimized for specific levels of average and peak powers. To improve the thermal design of microwave power transistors, total active area of the device is divided into a number of cells, which are distributed throughout the chip. In order to divide the area into cells, it is necessary to ensure that current, and therefore power, is equally distributed within the active area. Concentration of current in small sectors of the total chip area results in inefficient operation as well as fragile devices. The ballast techniques described in section 2.2.2 are used to improve the uniformity of the current distribution.

Microwave transistor cell design is dependent on the required operating conditions. Short bursts of RF power result in minimal junction heating, and, therefore, the performance of microwave transistors under short pulse, low duty factor pulsing primarily depends on the current and voltage handling characteristics of the geometry and diffusions. If duty factor remains low and pulsewidths are increased, peak-junction temperature rises during the pulse and results in decreased reliability. Eventually, if peak temperature is not limited, catastrophic failure may occur. In the case of short-pulse operation, as duty factor increases, the increase in average power dissipation causes an increase in junction temperature, which results in decreased reliability. Eventually, as the duty factor approaches continuous operation (CW), an extreme rise in junction temperature may also result in catastrophic failure. The rise in junction temperature is governed by the overall thermal design of the transistor, which includes the chip layout as well as package design. Increases in average-to-peak power ratio capability can be attained by designing a transistor that has its active area divided into small, thermally isolated cells. The trade-off involved in achieving better thermal isolation consists of requiring a physically larger chip. The larger chip results in higher chip costs because of the lower yield per wafer. It can also result in higher package costs if a larger package size is required. Because of these trade-offs, microwave transistor manufacturers tend to orient their device designs toward the requirements of various radar systems. Generally, it is optimal to minimize device size within the constraints of supplying the required power, gain, and system temperature. Transistor manufacturers generally classify radar pulse requirements according to the criteria given in Table 2-1.

Short-pulse systems allow for transistor chip designs with very dense active areas. High ratios of peak-to-average power can be attained. Very high peak-power transistors are generally manufactured for systems requiring low duty factors, such as IFF and DME systems. Most solid-state surveillance radars have either medium- or long-pulse requirements. These devices have somewhat lower peak-power outputs and a lower ratio of peak-to-average power capability. Some systems, especially those used in radar fuses, require high duty factors, and, therefore, can be approximated by CW requirements. Although they are approximated by CW limits, at 50 percent duty factor, transistors usually can provide 1 to 2 dB higher peak output for short-pulse systems than for CW operation. Systems having pulse requirements that fall in between these categories would require slightly different designs in order to achieve optimal performance. Generally, if average power-on-target is a prime consideration, bipolar transistors provide higher average power as the system approaches CW.

Table 2-1

Type of Device	Pulse Requirements
Short Pulse	Pulsewidth < 10 μs and Duty < 10%
Medium Pulse	Pulsewidth < 150 μs and Duty < 10%
Long Pulse	Pulsewidth < 10 ms and Duty < 25%
CW	Pulsewidth > 10 ms or Duty > 40%

2.2.5 DC Characteristics

Transistors used in radar systems are characterized by their dc breakdown voltage, leakage currents, dc current gain, saturation voltages, and static junction capacitances. Most radar systems use class C transistors biased with the emitter-to-base junction shorted by an RF choke. Under these conditions, the most critical breakdown voltage is that of the collector-base junction with the emitter-base junction shorted (BV_{CES}). During normal operation into a matched load, the peak collector-to-base voltage is approximately double the supply voltage. It is, therefore, normal to specify this breakdown voltage at more than double the supply voltage. As mentioned in section 2.2.1 on device geometries, most transistors use some degree of collector ballast, which effectively limits the collector current during breakdown. For this reason, short-pulse systems often specificy a breakdown voltage at less than double the supply voltage.

A less important breakdown voltage for radar transistors is the collector-emitter breakdown voltage with the base open (BV_{CEO}). It is specified primarily as a method of device processing control. Because the BV_{CEO} is inversely related to dc h_{FE}, it becomes a loosely specified parameter. For device control in radar applications, this parameter need not be specified.

The value of dc h_{FE} is also typically specified for radar transistors. Generally, dc h_{FE} is controlled by manufacturers in order to ensure proper gain and saturated power. It is, therefore, a result of proper wafer processing, which yields the required electrical performance, and hence it relates indirectly to device performance.

The collector-base junction capacitance (C_{ob}), is related to device processing, such as oxide thickness as well as collector epitaxial resistivity. C_{ob} affects device impedances and efficiency. Generally, it is optimal to keep collector-base junction and oxide capacitance as low as possible for a given amount of active area.

In addition to the above parameters, leakage currents are always specified for microwave power transistors. Leakage currents are measured under reverse bias of the transistor's junctions, and they are the result of defects in the silicon crystal or mobile ions trapped in the transistor's oxides. Generally, there are some small amounts of leakage currents present in high-power microwave transistors. Limits for leakage currents tend to be set at levels that allow reasonable transistor yields.

Because higher power transistors are effectively composed of paralleled junctions, the higher the power level, the higher the specified leakage. Unfortunately, the leakage may not be uniformly distributed between all parts of the active area; i.e., all of the leakage may result from current flowing in a single defect. In order to eliminate the lower reliability of devices having a large leakage source, high-current continuous breakdown tests may be used to cause unreliable leakage paths to melt and cause device rejection. High current BV_{CES} testing is an effective method of screening for such defects.

Additional dc characteristics of importance to microwave transistors include saturation voltages. The dc saturation voltages related to both the resistivity of the silicon used to manufacture the transistors as well as the electrical contacts to the junctions. Normally, saturation voltages are not specified for microwave power transistors because they are effectively screened by RF power tests. They can, however, be used to analyze devices with questionable performance. The dc characteristics are covered in more detail below.

2.2.6 RF Characteristics

The most important parameters of microwave power transistors, with respect to their operation in radar systems, are the RF characteristics. These parameters are defined and described below.

Power Gain is output power divided by input power, usually expressed in decibels. Power gain is most important at the anticipated RF operating point. Most solid-state radar systems operate with their transistor stages in the gain compression mode because the system then behaves more predictably over temperature and drive. *Power output with constant drive* is often specified instead of power gain. This is simply a different term for the same parameter.

Efficiency, usually called *collector efficiency,* is the RF output power divided by the dc input power and expressed as a percentage. Efficiency is important because it affects prime power requirements. Traditionally, microwave transistors have specified collector efficiency, so it is also important to have high power gain in order to minimize system power requirements. It is possible to specify a lower collector efficiency in conjunction with higher power gain, thereby obtaining higher system efficiency.

Bandwidth is usually specified by requiring that RF performance be measured at a number of frequencies in a fixed-tuned test fixture. In this case, the resulting bandwidth is not necessarily that of the transistor, but rather that of the transistor-circuit combination. Traditionally, radar systems require functional transistor testing (i.e., testing in a circuit derived from the system's circuit under system pulse conditions).

Stability is normally determined by observing the output spectrum of an operating transistor while varying drive, supply voltage, or load conditions to simulate anticipated system variations. Normally, stability refers to nonharmonic outputs that occur above the −60 to −80 dB level relative to the desired signal output. Stability testing is often difficult to perform because the test conditions are continuously varied.

Ruggedness is related to the ability of a transistor to withstand transient supply or load conditions without failure. Normally, a transistor is measured for performance under normal supply and load conditions, subjected to the transient conditions, and then retested under normal conditions. When retested, the device is considered acceptable when no measurable change in performance is observed.

Amplitude and Phase Characteristics are important in radar systems for a number of reasons. Perhaps the most obvious is because the power output of transistors is combined either within the system or in space (i.e., phased arrays). Power combining depends on the degree of match between the amplitude and phase characteristics of all transistors.

The primary amplitude specifications are those of *pulse droop* and *pulse rise and fall times.* Pulse droop is the ratio of the output power at the end of a pulse compared to that at the beginning of the pulse, usually expressed in decibels. Pulse rise and fall times are the measured periods taken for a detected pulse (envelope of the RF pulse) to rise or fall between the 10 percent and 90 percent amplitude points.

Two phase characteristics are important to radar systems: *insertion phase* and *phase settling.* Insertion phase is the electrical length of a transistor measured in degrees at a given test frequency. Phase settling is a measure of the length of time taken for the insertion phase to settle to within a certain limit of its steady-state (sinusoidal) condition.

Usually, only pulse droop and insertion phase are specified for radar system transistors. All transient pulse parameters depend on power supply conditioning circuits and energy storage systems, which are often too difficult to simulate during component testing. Other amplitude and phase characteristics, such as phase jitter, cannot be accurately measured on individual devices. These parameters are assumed to be sufficiently uniform after a device is qualified.

Junction Temperature is normally specified under simulated system test conditions for microwave power transistors used in radar system. Junction temperature is measured by using an infrared microscope to scan the junction under worst-case system conditions. Normally, the peak junction temperature is limited by military reliability requirements. Uniform junction temperature within the active area may also be a condition for acceptance. Because thermal scanning requires open transistor packages, junction temperature measurements are normally either a requirement for qualification or required on a sample basis.

2.3 TRANSISTOR PACKAGING

The previous section described transistor chips used in microwave power transistors and their electrical characteristics. Performance characteristics depend on the transistor package as well as the semiconductor chips. As mentioned earlier, modern microwave power transistors used in radar systems are actually hybrid devices consisting of active chips assembled in packages with passive matching elements. The package serves as the interface for the active silicon chips with the electrical and thermal design of a radar system. After the semiconductor chip, the package is the second most important element in determining or limiting transistor performance.

The major package considerations for radar transistors are electrical parasitics, thermal properties, mechanical properties, and environmental properties. These areas are discussed below.

2.3.1 Package Parasitics

Microwave transistor packages serve as the electrical, thermal, and mechanical interface for active semiconductor chip to the radar circuits. High power solid-state radar systems use a construction based on the repeated use of standard RF modules as building blocks. Packages used to contain RF power transistors influence the design of the radar modules. All solid-state radar systems currently in production use microstrip-type construction techniques because of performance and cost advantages. Therefore, microstrip or stripline packages are favored for the transistors used in these systems. Some of the characteristics of microwave transistors that dictate packaging requirements are as follows:

1. Die impedances require that the packages contain internal matching networks for all but the lowest power stages.
2. Thermal and electrical requirements dictate that transistor chips be soldered into their packages in order to obtain maximum thermal and electrical conductance.
3. Because the solder interface of the chips is made to the collector, packages must provide electrical isolation with a thermally conductive material.
4. Reliability requirements dictate that hermetic (preferably solder-sealed) packages be used, and that device assembly be made using an all-gold interconnecting system.
5. Most assembly techniques favor a package that is mounted with screws, although some assemblies use packages which are soldered directly into modules.

RF power transistor packages are constructed using a beryllia (BeO) ceramic base, which is metalized to form an isolated die attachment area for the transistor chip surrounded by a ground plane area, which allows mounting of matching elements (primarily MOS capacitors). An alumina (Al_2O_3) ceramic frame is usually attached to the BeO base, and serves to provide isolated hermetic feed-throughs for the transistor's input and output leads. The alumina frame is brazed to the beryllia base, and its top surface is metalized to allow solder attachment of a metal lid, which is used to hermetically seal the package. The beryllia ceramic usually has its ground plane connected by wrap-around metalization to the bottom of the base. This metalization allows for the electrical, mechanical, and thermal interface of the package to either a metal flange or the metal ground plane of a radar module.

A typical package assembly contains the active transistor chips plus internal input and output matching networks. Input matching networks usually consist of lowpass impedance-matching transformers designed by using the series inductance of bond-wires and the capacitance of shunt MOS capacitors soldered to the metalized ground plane. Output matching, if used, normally consists of shunt inductive bond-wires connected from the isolated collector-die attachment area to dc blocking capacitors mounted on the metalized ground plane. A photograph of a typical microwave power transistor is shown in Figure 2-7, and a schematic of this device is shown in Figure 2-8. The entire assembly is a hybrid circuit designed to match the transistor chip to external module impedance levels. Normally, package parasitics are incorporated in the matching networks.

Figure 2-7 Photograph of Pave Paws Transistor

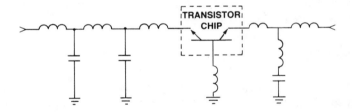

Figure 2-8 Schematic of Pave Paws Transistor

2.3.2 Thermal Properties

The thermal properties of the package are designed to complement those of the transistor die. Because it is required to match the thermal expansion coefficient of the silicon chips with that of the material on which they are mounted, most microwave power chips are mounted to metalized beryllia. At the time of this writing, other materials have been considered (such as diamond), but no material is used in volume production other than beryllia. Maximizing thermal transfer involves a number of trade-offs. Generally, more thermally conductive materials could be used at the expense of matching the materials' expansion coefficients. If the expansion coefficients are not well matched, package integrity is compromised over temperature. Thinner materials could be used, but again at the expense of mechanical integrity. Packages currently in use in solid-state radar systems generally use beryllia bases, which are between 25 and 60 mils in thickness. The metalized beryllia is usually soldered to either a copper, elkonite (a copper-tungsten alloy), or molybdenum flange, or to a module ground plane. Material thicknesses are optimized to reduce overall thermal resistance consistent with mechanical stress requirements.

2.3.3 Mechanical Considerations

In addition to the considerations mentioned above, microwave power transistor packages have other mechanical requirements. Generally, packages have stripline input and output leads. Leads are brazed to narrow ledges in order to minimize package parasitics. Lead strength is improved through the proper choice of a flexible lead material. Additionally, "*tee*" connections are normally made to the package in order to improve mechanical strength. Often, slotted leads are used because they result in a stronger lead braze and more flexure. Finally, packages must facilitate a hermetic sealing process.

2.3.4 Environmental Considerations

Most microwave power transistors are sealed with a high temperature gold-tin alloy solder. This is accomplished by soldering a gold-plated metal cap to a gold-plated ceramic seal ring in an inert atmosphere. The entire package must be capable of withstanding the assembly process, which subjects the packages to temperatures in the range of -55 to $+450$ degrees centigrade.

2.4 PEAK POWER *VERSUS* AVERAGE POWER

The power output of microwave transistors is limited by a number of parameters, but they can be generally classed as either electrical limitations or thermal limitations. These limiting factors will be discussed in the paragraphs that follow.

Electrically, the maximum output power that can be generated by a microwave transistor is determined by the maximum voltage at which the transistor can be operated times the maximum current that can be

conducted before saturation occurs times the maximum collector efficiency which occurs at an optimum load impedance. This can be written algebraically as

$$P_{Omax} = V_{CCmax} \times I_{Cmax} \times N_{Cmax} \qquad (2\text{-}1)$$

There is nothing profound here; this all sounds logical. The maximum supply voltage at which a transistor can operate is determined by the breakdown voltage of its collector-base junction and the expected peak voltage which will occur during the peak collector voltage swing. This peak voltage is approximately double the supply voltage. In an earlier section it was mentioned that due to collector ballast, the collector-base breakdown voltage (BV_{CES}) has a dynamic resistance characteristic in the breakdown region. This is illustrated by the transistor curve tracer plot of a typical collector-base breakdown shown in Figure 2-9. BV_{CES} is normally specified at relatively low current levels, and, therefore, it is possible to operate transistors above the one-half BV_{CES} level as long as the dissipation during breakdown is below the level which would cause secondary breakdown in the transistor. As a rule of thumb, for short, low duty pulsing (<1µs at 1%), the collector supply voltage can be run at 60 to 70 percent of the low current breakdown voltage.

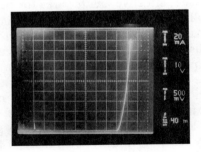

Figure 2-9 Photograph of Typical BV_{CES} Showing Dynamic Resistance in the Breakdown Region

The amount of collector current that a microwave transistor can conduct is determined by its *effective emitter area*. Effective emitter area refers to that area which is actually used, and it is less than the total emitter area because of various conduction-limiting phenomina, such as base widening and current crowding. This was discussed in an earlier section, and it was pointed out that microwave transistor designs generally maximize the emitter periphery-to-base area ratio, so that they can improve emitter current injection. High output power is obtained from microwave transistors when they operate in the class C mode. In this mode, current is conducted for less than 180 degrees of the RF cycle. Therefore, the average collector current is less than half of the peak value.

Collector efficiency depends on a number of things, including saturation voltages, charge storage, parametric capacitor losses, and $f(t)$. Collector efficiency is inversely proportional to saturation voltages and directly proportional to $f(t)$. An ideal transistor operated as a class C amplifier would conduct an impulse of current during the "on" portion of the cycle, and would have zero voltage drop during that time of conduction. Real transistors conduct limited current, have a voltage drop during current conduction, and require a finite time to turn on and off. Normally, maximum output power does not occur at the operating point where maximum collector efficiency occurs. Transistors produce their maximum output when they are driven into hard saturation. This causes higher saturation voltages and longer conduction angles, but it also results in higher output power.

To sum up the foregoing discussion, peak power output is primarily limited by parameters related to the transistor's geometry and processing. These parameters include breakdown voltage and collector ballast, which limit operating voltage; effective emitter area (or periphery) and current sharing (emitter ballast), which limit current handling capability; and $f(t)$ and saturation voltage, which limit efficiency under saturated operation.

Thus far, this discussion has been limited to peak output power obtained under short pulse, low duty cycle conditions. During an RF pulse, heating occurs due to power dissipated in the transistor. This heating causes the saturation voltage to rise and carrier mobility to fall, and output power sags as pulsewidths are increased. If sufficient power dissipation occurs, a localized melt can result in the semiconductor chip. Local melts usually result in catastrophic failure of the transistor.

The thermal limitations of a microwave transistor depend on its thermal design. Heat generated in the junctions must be conducted to a heat sink by way of a path consisting of the silicon die, the transistor package plus materials used to attach the die to the package, the amplifier housing, *et cetera*. It is common for manufacturers to characterize their transistors on a standalone basis by referring to the junction-to-case thermal resistance (θ_{jc}). System designers can then use this thermal resistance plus the thermal characteristics of their system to determine the expected peak junction temperature for the system.

Thermal resistance is obtained by measuring a transistor's junction and case temperatures under RF operation and calculating the dissipated power. Thermal resistance is then defined as

$$\theta_{jc} = (T_j - T_c)/P_{diss} \tag{2-2}$$

where

$$P_{diss} = P_{dc} + P_{in} - P_{out}$$

In order to be a valid system calculation, the thermal resistance of the transistor must be measured under system pulsewidth and duty cycle conditions. Because the thermal resistance is a function of temperature, it should be measured at the maximum case temperature anticipated for the system. This last factor can be ignored if an estimated correction factor is added to the value determined under standard conditions (flange temperature of 25 degrees centigrade is usually standard). When a system operates with a case temperature of 100 degrees centigrade, the correction factor causes thermal resistance to increase approximately 15 percent. This can often be ignored.

A transistor's thermal resistance is determined by the path between the junction and the package case, and, therefore, depends on the package materials and physical layout. Figure 2-10 illustrates the cross section of a typical microwave transistor. Heat flows from the junction through the silicon, die attachment solder, package metalization and ceramic, package braze or solder, and, finally, the flange. The heat spreads as it flows. If materials with relatively high thermal conductivities are used in the path of heat flow, most of the heat can be considered to flow within a spreading angle, as depicted in Figure 2-10. Figure 2-11 shows details of the heat path for the transistor die with its active cells. Heat from electrically isolated cells overlaps at a specific point in the vertical structure, depending on the spacing between the cells. Wider cell spacing reduces the overlap and, therefore, reduces thermal resistance. The price paid for lower thermal resistance is a lower electrical density on the silicon chip. This translates into larger chips for a given power level. Larger chips have lower die count on a processed wafer and require larger packages with higher package parasitics. Both lower die count per wafer and larger packages result in higher transistor costs.

These are some of the trade-offs involved in the design of a power transistor chip. Short pulse, low duty cycle transistors can be designed with high densities, making it cost effective to design high power devices. With low average power, the thermal design is not usually critical. As operation approaches CW, thermal design becomes very critical and active cells are spread so as to reduce thermal resistance.

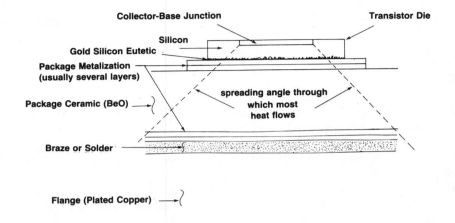

Figure 2-10 Cross Section of a Typical Microwave Transistor Assembly Showing Heat Flow

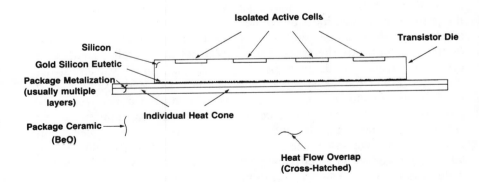

Figure 2-11 Heat Path in Cellular Transistor Showing Overlap of Thermal Paths

When transistors are pulsed in a radar system, the junction heats with each pulse. Transient thermal characteristics depend on thermal resistance, as discussed above, plus the heat capacity of various materials used in the transistor assembly. Transient response has a number of time constants corresponding to the thermal resistance and heat capacity of various materials comprising the packaged transistor. This is illustrated in Figure 2-12 which shows thermal resistance *versus* pulsewidth for a single-pulse system.

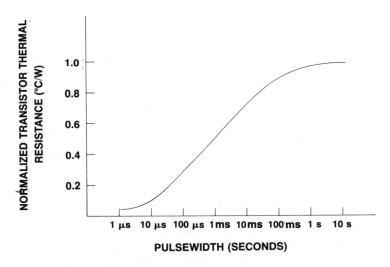

Figure 2-12 Normalized Transistor Thermal Resistance (Θ_{jc}) Illustrating Transient Response *versus* Pulsewidth for Single-Pulse (Low Duty Cycle) Systems

Transistors can be designed to favor short-pulse systems, but high average power favors wide pulsewidths at high duty cycle. Most transistors produce their highest average power under the limiting conditions of CW operation. This is illustrated in the graph of Figure 2-13, which plots average power *versus* pulsewidth and duty cycle where junction and case temperatures are kept constant.

In summary, peak power output of microwave transistors is limited by electrical conditions while average power depends more heavily on the thermal design of the transistor. Transistors can be designed to optimize either short pulse, low duty cycle operation, or long pulse, high duty cycle operation. In general, high average power favors long-pulse operation.

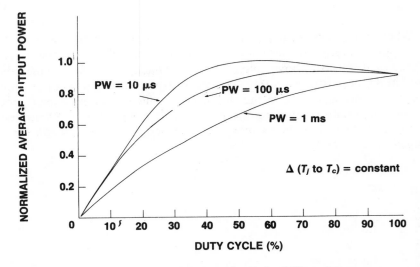

Figure 2-13 Average Output Power of a Pulsed Transistor *versus* Pulsewidth and Duty Cycle with Case and Junction Temperatures Constant
Note: Peak power capability is limited for very narrow, low duty cycle pulses by the transistor design

2.5 RELIABILITY CONSIDERATIONS

The reliability of microwave power transistors is often discussed and quoted, but it is not well supported by experimental evidence. This section will give an overview of device reliability based on the experience of various manufacturers of microwave power transistors with respect to several known failure mechanisms. Part of the reliability problem is that the statistical nature of failure mechanisms requires large quantities of data taken over time to improve the forecast. This discussion extends beyond the reliability considerations normally reported. The hope is that solid-state radars being manufactured today will form the test bed for establishing more thorough and accurate reliability predictions in the future.

Failure mechanisms for microwave power transistor can be classified as follows:

1. Metalization failures;
2. Bond failures;
3. Capacitor failures;
4. Package failures;
5. Mechanical stress failures;
6. Electrical stress failures.

2.5.1 Metalization

Metalization failures involve the most widely discussed failure mechanisms of microwave power transistors. A survey of the literature on microwave power transistor reliability would conclude that this is the *only* area of questionable reliability. In fact, because it has been so widely studied, it is probably the *least* important area for today's transistors.

Several types of metalization failures can occur, of which metal *migration* is probably the best understood. This mechanism is caused by high current density in the metalization stripes of the semiconductor chips. It results in mass movement of metal ions due to the combination of metal ion kinetic energy at higher temperatures in conjunction with momentum interchange between the electrons flowing in the conductors and the metal ions. Metal ions which are displaced within the conductors form wiskers on the metal surface and cause voids to be formed. Eventually, the voids result in an open circuit. The degree of voiding depends on time, temperature, current density, and the activation energy of the metalization system. This last factor depends mostly on the type of metal. Early microwave transistors used aluminum for the primary conductor, which resulted in unacceptable metal migration at relatively low current densities and temperatures. Modern microwave transistors use gold metalization, which has resulted in minimal migration because of the higher activation energy of the gold system. Additionally, modern microwave transistors generally have lower designed current densities (about $5 \cdot 10^5$ A/cm^2), which also results in improved resistance to migration.

A second important metalization failure mechanism is caused by *diffusion* of metal into the silicon, causing a short in the shallow emitter-base junction. Because gold alloys with silicon at low temperatures, a barrier is placed between them. Breaks in the barrier or diffusion of gold through the barrier can cause junction shorting. The type of barrier, method of processing the transistor, design of the transistor geometry, and even the processing used for specific wafers can all have an effect on the resistance to gold alloying. Resistance to gold alloying can be designed into a microwave transistor by proper barrier selection, junction passivation, contact window design, *et cetera*. The tendency to alloy can be tested by screening transistors at temperatures above that which causes alloying (i.e., above 420 degrees centigrade).

A third prevalent failure mechanism is *peeling* of the metalization. In this case, metal fails to stick to either the silicon or the oxide which passivates the transistor. Stresses in the assembly process or within a finished transistor can pull the metal from the chip, resulting in an open circuit. Usually, nondestructive bond-pull testing can be done to eliminate these types of failures.

2.5.2 Bonds

Bond failures can result in two principal conditions: open circuits and short circuits. It was mentioned previously that older aluminum transistors have been known to fail due to bond-wire fatigue resulting from pulsed operation in radar systems. This type of failure can occur in other metal systems. Not much is known about such failures because of the lack of information on the relationship between metals, wire currents, temperatures, pulse characteristics, and failure rates. Specific testing must be done in radar systems to determine the fatigue characteristics of the transistor bond-wires under various pulsing conditions.

More common bond failures are the result of bonds lifting from either die bonding pads or package metalization. Bonds that lift from die pads are either the result of improper bonding machine operation or contamination of the metalization surface which is to be bonded. Machine settings are normally checked periodically to ensure proper operation by performing destructive bond-pull tests on transistor samples. In addition to ensuring proper machine settings, these tests may also isolate potential contamination problems. Additional problems can occur due to bonding on the eutectic solder of the package metalization. The eutectic is usually a result of die attachment operations. Wires bonded on eutectic result in failure over time. Bond strength tests performed over time using high temperature aging (200 degrees centigrade) to accelerate the failure mechanism have shown that bonds on eutectic suffer from a significant reduction in bond adhesion, which eventually results in failure. Most radar systems use transistors which are visually screened to eliminate bonds on eutectic. Additionally, testing can be used to evaluate the effects of bonding on contaminated bonding pads. Normally, this testing consists of performing sample wire-pull tests before and after a period of high temperature storage. Degradation in bond strength is a sign of pending failure.

In addition to breaking or forming open circuits, bonds can short to various elements within the package. Commonly, shorts occur either to the edge of semiconductor chips or to closely spaced emitter and base metal fingers on the transistor chips. Shorts to the edge of dice can be eliminated by proper bonding techniques, and they can be screened by visual inspection. Bonds that short metal fingers of the dice can be reduced by passivating the die metalization. Additionally, proper bonding and inspection can eliminate this type of short circuit.

2.5.3 Capacitors

Most microwave power transistors contain MOS capacitors, which are used as internal matching elements. Capacitors are most likely to fail because of either cracking or oxide defects. Cracking will be treated later under mechanical stress failures (section 2.5.4). Oxide defects can result in pinholes in the dielectric that eventually cause shorts. When matching capacitor chips are tested at the wafer level, they are commonly screened for leakage current. This is a gated-pulse test. If the capacitor has a small pinhole, the testing can often burn out the short before the gated leakage measurement is completed. Such a capacitor then passes testing and is used in a transistor assembly. Pinholes are normally very difficult to spot visually and, therefore, transistors having capacitors with pinholes are likely to be shipped as good transistors. Under continuous operation, however, the pinholes can result in a short, which causes device failure. This is more common in capacitors used for the output matching of transistors because of higher applied voltage as well as higher power. Special breakdown testing of capacitors at the wafer level, or leakage testing after capacitors are mounted in packages, can be used to eliminate a large percentage of capacitor breakdown failures.

2.5.4 Package Failures

Generally, packages are involved in the highest percentage of all microwave power transistor failures. Some of the "package" failures are the result of improper use, but the end failure can only be attributed to the package. Package failures can be classed as follows:

1. Lead failures
2. Package cracking
3. Package contamination

Lead failures can viewed as resulting either from excess stress applied during transistor manufacture, to circuit fabrication, or to poor lead design and fabrication. If a package can only withstand two pounds of force and it experiences three pounds of force during assembly, test, or installation, it is likely to fail. Microwave package leads are designed to have minimum parasitics. This generally results in a relatively fragile lead-to-package attachment. Improper transistor assembly equipment, improper handling, or poorly designed test fixtures can stress and, therefore, weaken the lead-to-package interface. Radar circuits which do not properly interface to leads by taking into account lead height and the match of thermal coefficients of expansion can pull the leads from even the best packages. Reliable solid-state radars must establish proper screens for lead adhesion, ensure that radar circuits have proper mechanical design, and guarantee that handling during assembly does not excessively stress the leads. Modern microwave power transistor packages have incorporated a number of design improvements such as *"tee"* leads, which result in a reduction in lead failures.

Package cracks often propagate to MOS capacitors or the transistor chips, and thus cause failure. Additionally, cracked packages may allow the progress of contamination and moisture, which eventually cause failure. Package cracking often results from improper installation of transistors in amplifier circuits. Cracking also occurs because of poor package design or assembly. Ceramic microwave packages are made of materials having different coefficients of thermal resistance. These differences must be taken into account to produce transistors that will not crack under "normal" mechanical and thermal stress. Thermal and mechanical screening must be used to qualify reliable package assemblies. Radar assemblies must ensure that transistors are properly attached to smooth, flat surfaces, so that the ceramics are not overly stressed to the point of causing cracking.

2.6 TYPICAL PERFORMANCE OF SOLID-STATE DEVICES

Up to this point, various aspects of RF power transistors have been discussed in general terms without reference to actual device performance that has been either reported, specified, or advertised. Solid-state radar systems represent a major market for silicon RF power transistors. Typically, transistors are designed to meet the requirements of specific radar systems within the limits of device producibility and cost. For this reason, gaps in transistor performance tend to form between the frequency, power, and pulse requirements of major solid-state radar systems. Performance referred to in this section should be referenced to the period in which the material was written (*circa* 1985). Anticipated advancements in performance will be discussed later.

In order to understand present transistor performance, it is important to become familiar with some general characteristics of transistor design and construction. This material was covered earlier and a brief summary follows.

For a given transistor design, both efficiency and gain decrease as frequency increases. Conversely, for a given design, ruggedness (the ability to withstand over-drive, over-voltage, and load mismatch) increases as frequency is increased. Because of this divergence between electrical performance and ruggedness, transistor chips are normally designed for optimal operations over a limited frequency range.

High power RF transistors require low reactance between their common terminal (emitter or base) and ground. Therefore, the number of ground bond-wires relative to a given amount of power (impedance) must increase as frequency increases. For low-frequency operation, transistors can be constructed with large active cells, whereas higher frequency operation forces the design of smaller, lower power cells. An alternative approach at higher frequencies is to incorporate multiple grounding pads on somewhat larger cells. In any case, for a given level of output power, higher frequency transistors require more ground bond-wires than lower frequency transistors.

Cell design was discussed earlier. The main criteria for selecting a cell design (aspect ratio and cell spacing) are related to thermal requirements. Narrow pulse, low duty factor operation allows for designs having a dense active area and, therefore, relatively high dc thermal resistance. Current handling capability of both the metal stripes and bond-wires may still limit cell density. Also, distribution of the ground and input bonding pads must be considered to minimize inductance as mentioned above.

Manufacturers must always consider processing and assembly yields when selecting chip and transistor designs. Large chips exhibit a high probability of having a defect within their active area, thereby causing rejection. Because of yield considerations, manufacturers often choose to divide the active area of higher power transistors into smaller chips and place multiple chips into a single package assembly. The multiple chips mounted within a package are matched by selecting them to be adjacent on a processed wafer. Selecting adjacent chips ensures uniform operation of the assembly because the chips have approximatly the same gain and impedance. Use of multiple chips improves yield for higher power transistors within the limit of obtaining a good die attachment for all of the chips. Eventually, the use of a large number of chips placed into a single package is limited by assembly yields.

Obviously, device configuration influences the performance of transistors. Most devices used at the higher microwave frequencies (over 1 GHz), are configured as single-ended, common-base transistors. Lower frequencies tend to favor the common-emitter configuration. A number of lower frequency radar systems use push-pull transistors in their high-power transmitters. A push-pull transistor effectively places two transistors within a single package. It offers advantages over using two discrete transistors insofar as manufacturers match the chips within the package.

A virtual ground is also formed withing the package, which yields lower parasitic inductance than could be realized if separately packaged transistors were used. Push-pull is not generally used at higher frequencies because of stability problems. Push-pull operation offers advantages in terms of device power and impedances at the expense of requiring somewhat more complex balanced matching circuitry.

At the time of this writing, most solid-state radars operating at or below S band that are in the engineering design model or production phases use silicon bipolar devices. Exceptions may occur for relatively low power radars, which may use GaAs FETs or two-terminal devices. Silicon bipolar transistors are preferred because of their performance, price, and reliability history. The most prominent alternative technology today is the RF vertical MOS FET (VMOS). While significant improvements have been made in FET technology, performance and reliability history favor the use of bipolar devices. It is probable that VMOS FETs, or other alternative technical devices, will be utilized in radar systems in the future. Alternative technical devices are discussed later in this chapter.

2.6.1 Power Output, Gain, and Efficiency *versus* Frequency

As noted above, narrow pulse, low duty factor operation allows manufacturers to design very dense active areas, whereas long pulse, CW operation causes greater spreading of the active area. In reviewing transistor performance, it is important to relate that performance to pulsing requirements. This section presents typical RF performance of modern transistors operating under various pulse conditions across the frequency range from HF through S band. Performance is presented in a series of graphs divided along the lines of pulse requirements. Pulsing is divided generally into the categories used earlier in this chapter: short pulse, medium pulse, long pulse, and CW. Figures 2-14 and 2-15 present curves showing output capability of power transistors *versus* frequency for the various pulse conditions. The curves are fitted to data extracted from reported performance of various radar systems in either the production or design phase. Data points are included for reference, and symbols indicate whether the performance represents production transistors or development devices. Figure 2-16 shows the gain *versus* frequency for transistors used in output stages of various radar systems. A range of gain is shown. Generally, higher gain could be expected for narrow pulse, or lower power, operation. Finally, collector efficiency range *versus* frequency is shown in Figure 2-17. The efficiency shown is representative of higher power transistors typically having powers and gains shown in

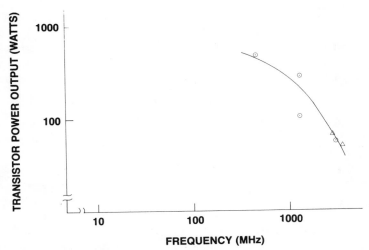

Figure 2-14 Transistor Power Output *versus* Frequency for Short and Medium Pulse Solid-State Radar Systems

Note: ⊙ indicates that system is in production
△ indicates that system is in development

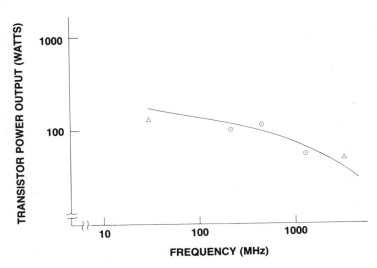

Figure 2-15 Transistor Power Output *versus* Frequency for Long Pulse and CW Radar Systems

Note: ⊙ indicates that system is in production
△ indicates that system is in development

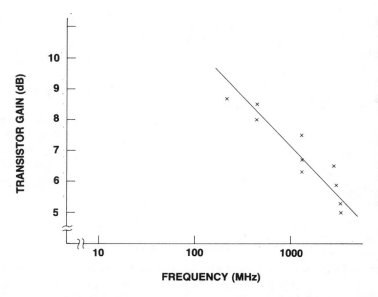

Figure 2-16 Range of Gain *versus* Frequency for Transistors Used in the Output of Radar Transmitters

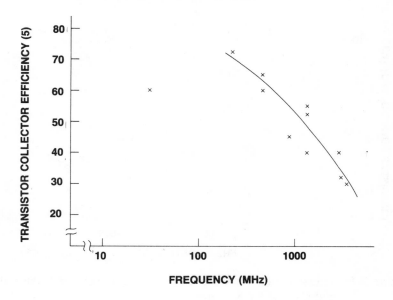

Figure 2-17 Range of Collector Efficiency *versus* Frequency for Transistors Used in the Output of Radar Transmitters

Figures 2-14 through 2-16. The data presented are interrelated, and it is possible to improve one parameter at the expense of others (i.e., improve efficiency at the expense of gain or power output).

The data presented in Figures 2-14 through 2-17 are interrelated and it should not be assumed that power output, gain, and efficiency can all be optimized at the same time. Table 2-2 shows a summary of the performance of output transistors used in various solid-state radar systems. The data can be used to relate the various performance parameters.

Table 2-2 Summary of the Performance of RF Power Transistors Used in Solid-State Radars

Radar System System	Frequency (MHz)	Pulse	Transistor Performance P(O) (W)	Gain (dB)	Collector Efficiency (percent)
OTH	5 to 30	CW	130	14.0	60
NAVSPASUR	217	CW	100	8.7	72
AN/SPS-40	400 to 450	60 uS @ 2%	450	8.0	60
PAVE PAWS	420 to 450	16 mS @ 20%	115	8.5	65
AN/FPS-50 (BMEWS)	--- Same as PAVE PAWS ---				
AN/SPS-49	L-BAND	250 uS @ 10%	105	6.7	45
AN/TPS-59	1235 to 1365	2 mS @ 20%	55	6.6	52
SEEK IGLOO and NORTH WARNING	--- Same as AN/TPS-59 ---				
RAMP	1250 to 1350	100uS @ 10%	105	7.5	55
MARTELLO S723	1235 to 1365	150uS @ 4%	275	6.3	40
MATCALS	2700 to 2900	100uS @ 10%	63	6.5	40
AN/SPS-48	2900 to 3100	40uS @ 4%	55	5.9	32
AN/TPQ-37	3100 to 3500	100uS @ 25%	30	5.0	30
HADR	3100 to 3500	800uS @ 23%	50	5.3	35

2.6.2 Bandwidth Limitations

As might be expected, transistor performance is related to system bandwidth. In narrowband systems, i.e., those having a bandwidth of up to a few percent, transistors can be operated at high voltages and matching can be nearly optimal across the band. Optimal matching produces optimal performance. As bandwidth expands, two important factors affect and limit performance. These are discussed below.

Transistor parasitic reactances were discussed earlier. The ratio of resistance to reactance in a parallel circuit, referred to as Q, limits the bandwidth capability of a transistor. As transistors are matched over wider bandwidths, the degree of mismatch within the band increases. This mismatch reduces electrical performance in terms of power output, gain, and efficiency. Wideband radar systems, therefore, compromise device performance.

When wider bandwidths are required, transistor manufacturers must take the bandwidth into account in the design of their components.

Normally, peak output powers can be optimized by using relatively high supply voltages. To adjust for the higher system voltages, processing must be targeted for higher semiconductor breakdown voltages. As supply voltage is increased, transistor load resistance (across the transistors intrinsic current generator) goes up in proportion to the square of the voltage. On the other hand, transistor junction capacitance decreases relatively linearly with increasing voltage. The ratio of the resistance to reactance or output Q, therefore, increases directly with increased design voltage. This increase in output Q limits the available bandwidth of the transistor. Stated differently, wideband operation favors lower transistor operating voltage, and the lower voltage, in turn, limits peak power capability.

2.6.3 Stability and Ruggedness

As noted earlier, transistor design and processing trade-offs allow some performance parameters to be optimized at the expense of others. Stability of a transistor refers to its ability to operate without generating spurious output signals. Obviously, transistors must be stable when operating into a normal load impedance. The ability to operate into mismatched loads without producing spurious outputs is a measure of a transistor's stability. Stability is not independent of matching circuitry. Circuits can enhance stability by virtue of their out-of-band impedance characteristics, their input-to-output isolation, and their bias decoupling. Generally, it is not necessary for radar amplifiers to be unconditionally stable (i.e., stable with any combination of source and load impedance). However, they should be sufficiently stable to produce a spurious-free output into all system loads.

In addition to stability, ruggedness is an important performance parameter. Ruggedness refers to a transistor's ability to withstand overdrive, supply transients, and load mismatch without failure (i.e., no degradation in performance when returned to normal operation). As with stability, a transistor must be able to withstand normal system stress. Margin in ruggedness can allow a transistor to withstand unusual and unanticipated manufacturing or operating stress.

Stability and ruggedness are somewhat related. If a pulsed transistor becomes unstable and oscillates, average power may increase to the point where the transistor fails. Improvements in stability will generally improve ruggedness.

Neither stability nor ruggedness are specifically taken into account in the performance data included in this section. As a general rule, all of the devices included in the performance section are sufficiently stable into a 2 to 1 load mismatch (in-band) and sufficiently rugged into a 3 to 1 load mismatch. Some of the devices will tolerate higher mismatch and meet stability or ruggedness testing. As independent parameters, stability can generally be enhanced by reducing device gain, and ruggedness can be improved through device ballast techniques, which will limit saturated power. Stability and ruggedness are discussed further in the next section.

2.7 SPECIFYING POWER TRANSISTORS FOR SOLID-STATE RADAR SYSTEMS

The first step in specifying transistors for solid-state radar systems is to verify that the transistors selected meet the overall system requirements of performance, environmental, and reliability. Devices often must be subjected to extensive evaluation, step stress testing, and destructive analysis to determine their limits of performance. Device performance should be guard-banded and, if possible, evaluated over time during the system design phase. Performance failures should be thoroughly analyzed. Solid-state radars use large numbers of transistors, and it is important to understand performance and reliability limits early in the selection process. Reliability must be designed into the transistors. Screening to a specification can only be used to weed out the few devices which fall outside normal performance or quality limits.

Performance specifications should focus on parameters with known relationships to the end use. Transistor specifications often include parameters which are not important to the end use while excluding others of importance. If second sources are important, the relationship between various parameters from different suppliers must be taken into account. In particular, dc parameters of transistors may differ significantly between various manufacturers, although RF parameters are nearly identical.

Specifications can be grouped into the following areas:

1. dc characteristics;
2. RF characteristics;
3. Mechanical characteristics;
4. *High Rel* processing requirements.

Each of these areas will be discussed below.

2.7.1 DC Characteristics

The dc characteristics are used to control the selection of RF power transistors because of their relationship to performance and reliability as well as their ease of measurement. Unfortunately, it is common to find that the dc parameters specified are often simply copied from a vendor's standard data sheet. The following categories of dc characteristics may be specified:

Breakdown Voltages. The dc breakdown voltages of transistors are specified primarily to ensure that these limits are not exceeded during operation. Most RF power transistors are operated in class C made with their emitter-to-base junction externally shorted through a low resistance dc return. From the standpoint of actual operation, the dc breakdown voltage of importance, then, is the collector-to-emitter breakdown with the base-to-emitter junction short circuited (BV_{CES}). If class AB or class A operation is intended, it may also be important to specify BV_{CER} or BV_{CEO}, respectively, the collector-to-emitter breakdown voltage with the base-to-emitter interconnected with a resistor or left open. The idea is to specify a breakdown which is consistent with actual use.

The specification limit for collector-to-base, or collector-to-emitter, breakdown voltage should be based on system requirements. Under class C operation, the peak collector voltage is theoretically approximately double the supply voltage. Normally, breakdown voltages are specified at a current level that is 5 to 10 times the maximum leakage current at nominal operating voltage (*see* leakage currents below). As pointed out earlier, intrinsic collector ballast results in significant dynamic resistance in the collector breakdown region. For this reason, it is typical to operate short pulsed RF power transistors above the traditional one-half BV_{CES} level. Normally, the breakdown specification is 1.5 to 2.5 times the collector operating voltage for class C operation. This can be used as a rule of thumb. The exact value should be selected on the basis of proven reliability of a transistor at the voltage specified under anticipated load and power supply transient limits.

For class A operation, peak collector voltage depends on the type of collector load coupling. A purely resistive load (dc and RF) results in a peak voltage equal to the collector supply voltage. The use of transformer coupling can result in a peak collector voltage of up to two times the supply voltage. Actual operation must be analyzed to determine breakdown voltage requirements. Because class A operation can result in sustained peak voltages, it may be necessary to increase the breakdown specification in order to ensure reliable operation.

It is interesting to note the number of specifications that include a minimum emitter-to-base breakdown voltage (BV_{EBO}). As pointed out earlier, breakdown voltages depend on diffusion profiles and doping concentrations. Emitter-to-base breakdowns of bipolar RF power transistors normally fall in the range of 4 to 8 V. The only significance of the exact voltage is related to process control. The BV_{EBO} specified by most vendors is approximately 70 percent of the actual breakdown value, and it is common to find that this specification is used on source control documents. In fact, this specification is more of a leakage specification. There is no need to specify BV_{EBO} for most RF power transistors.

Leakage Currents. Perfect semiconductor crystals having well-isolated metalization and pure oxides have zero leakage currents. Leakage currents are the result of various imperfections within the transistor chips and packaging. Processing of RF power semiconductors can result in a leakages ranging from nearly zero current to almost perfect shorts.

There are three considerations involved in specifying leakage currents. First, some radar systems operate at very low duty cycles and may require low dc leakages to keep total system current at a low level. This is primarily important when the duty cycle is below 1 percent. Even higher duty cycle systems can have their efficiency measurably degraded by allowing liberal leakage currents. When duty cycle exceeds 10 percent, however, the degradation in overall efficiency due to leakage current is not usually important.

The second aspect of leakage currents deals with reliability. Because leakage currents are the result of defects, leakages should be specified so that unreliable sources of leakage are eliminated. Unfortunately, machines used to screen leakage currents cannot normally differentiate between the various sources of leakages. (Sometimes, high current breakdown testing can, however, as will be covered in a later section). One way to try to limit unreliable defects is to measure the leakage current at a number of voltages and ensure that the currents, as a maximum, increase linearly. Leakage currents are normally specified at nominal operating voltage. A second specification can be added at between 1.3 and 1.7 times the nominal operating voltage with the limit increased proportionally. This does not mean that nonlinear leakage current *versus* voltage cannot be passed, but it does improve the leakage screening. Other leakage current screens which improve reliability can be added, and these are covered below.

The third criteria for setting leakage currents occurs when noise generation is important. Normally, this is not a factor for radar transmitter applications. In cases where power transistors are specified for oscillator applications, high leakage currents can result in an increased noise floor and measurements must be used to determine the importance of leakage currents on output noise floor.

DC Gain. The dc gain (h_{FE} for bipolar transistors; g_m for FETs) is another example of a parameter which is specified more because it appears on vendor data sheets than because it relates to operational requirements. An exception to this would be the case where class A or class AB operation is used. In this case, dc gain can affect bias circuit components.

As discussed above under emitter-to-base breakdown voltage, dc gain is the result of diffusions. Because the principal criteria for radar transistors are gain, efficiency, ruggedness, and reliability, dc gain is not usually important. An exception sometimes occurs when stability is related to dc gain, but stability can be better screened under RF testing. The dc gain is often used for reliability screening as will be discussed below.

As a guideline, if dc gain is specified, it is best to specify it in the region of anticipated operation. Because of measurement problems, it is common to specify h_{FE} at 5 V or lower. This is done in part because h_{FE} is a current parameter and, therefore, does not vary significantly with voltage.

2.7.2 RF Characteristics

Unlike dc parameters, RF characteristics directly relate to system operation. RF characteristics for power transistors used in radar systems are discussed below. Because radar systems represent high volume application of RF power transistors, it is common to require performance testing in specific circuits and under conditions that directly relate to the radar system requirements. Such functional testing is the heart of many procurement specifications. The following list includes parameters normally specified:

1. Operating frequency;
2. Output power;
3. Power Gain;
4. Efficiency;
5. Pulse fidelity;
6. Ruggedness;
7. Stability.

Operating frequency, power output, and gain are straightforward. As mentioned above, it is common to specify that these parameters be screened by testing in a fixture that is derived from system circuitry.

Efficiency as specified for most RF power transistors is output efficiency (collector or drain efficiency for bipolar transistors and FETs, respectively), which is defined as RF output power divided by dc input power. It is also possible to specify power-added efficiency, taking into account RF input power. This is somewhat redundant in that power gain essentially covers this. Efficiency specifications are based on system requirements as well as the ability of vendors to manufacture transistors that meet the specification with sufficient yield. This becomes something of a pricing function.

Pulse fidelity can be specified in a number of ways. Primarily, pulse fidelity is specified in terms of pulse rise time, fall time, and pulse droop. The main criteria for pulse fidelity relate to system performance requirements, including output spectrum. Pulse droop specifications may also be used indirectly to indicate, or limit, thermal characteristics of microwave transistors.

Ruggedness, or the ability of a transistor to withstand electrical stress, is usually specified by requiring testing with a combination of overvoltage, over-drive, and mismatched load conditions. There are two purposes for ruggedness testing. One is to ensure that no failure occurs under transient load conditions. Additionally, ruggedness may relate to radar manufacturing yield because it is common for module assembly and testing to overly stress components. Selection of ruggedness limits involves a trade-off between procurement costs *versus* system manufacturing yields and costs. Specification levels must be based on measurements and calculations of maximum stress anticipated in a radar system environment. Transistor manufacturers use processing to trade off ruggedness with other RF performance parameters, such as gain and efficiency. Specifications should be consistent with the state of the art for transistor manufacturing.

Stability relates to fidelity in that stable transistors have a spectral output, which is directly related to that of their input signal. Stability is normally specified under the range of stress which is expected under normal system operation. In this case, spectral purity is guaranteed under variations of drive, supply voltage, and load impedance.

2.7.3 Mechanical Characteristics

Mechanical characteristics of RF power transistors relate to their size limits as well as control of the materials used in their manufacturing process. Additionally, mechanical characteristics are specified as follows:

1. Lead plating and soldering ability;
2. Lead strength;
3. Package sealing;
4. Bond strength;
5. Environmental characteristics;
6. Shock and vibration;
7. Package marking.

The above characteristics are beyond the scope of this text and are adequately covered in various military handbooks, which should be consulted for further information.

2.7.4 *High Rel* Processing Requirements

Small radar system components such as RF transistors have amazingly long specifications. Not surprisingly, a large part of these specifications are devoted to special processing designed to ensure long-term reliable operation. *Rel* processing generally follows standard semiconductor screening. The dc characteristics were discussed above, and it was pointed out that leakage currents were primarily the result of imperfections in the semiconductor crystal. It is important to determine whether these imperfections (or defects) limit transistor reliability. Additionally, dc gain, while normally unimportant from a performance standpoint, is used as an indicator of transistor reliability. Two primary reliability screens which use dc characteristics as transistor reliability monitors are *burn-in* and *high temperature reverse bias* (HTRB). These screens are discussed below.

The term "dc burn-in" is exactly what its name implies. Transistors are burned in, or operated with dc bias, for a period of time (usually 168 hours) with a forced peak junction temperature. The dc gain and leakage parameters are measured before and after burn-in. The pre- and post-burn-in readings are compared. Ideally, these parameters would be monitored over time, and the measured values of leakage and gain would be plotted. If the parameters stabilize at acceptable levels, a reliable transistor can be predicted. Lack of stability in these parameters would indicate a suspect reliability. A much better screen would be an "RF burn-in," but the cost of RF burn-in has been assessed as too high in comparison to the value it might yield. Unfortunately, there is almost no correlation between the conditions used for dc burn-in compared to those which exist during actual operation. The process of dc burn-in attempts to isolate and reject transistors which have processing or material defects that cause junctions to degrade because of current flow or localized heating over time.

HTRB screening is performed by reverse biasing a transistor's collector-base junction, while heating the transistor for a period of time, usually 48 hours. The dc leakages, primarily, I_{CES}, I_{CBO}, or I_{CEO}, are monitored and compared before and after this processing. A significant shift in leakage current is considered to be a sign of contamination, which in turn indicates a suspect reliability. This screen primarily isolates anomalies associated with oxide defects or processing cleanliness. Defects uncovered by this screen can be caused by either wafer or assembly processing. Significant quantities of failures are typically grouped by either a particular wafer or a particular assembly lot. Sampling for *Rel* testing can be used to reject wafers with defects before adding the expense of assembling a large quantity of material. This is especially effective when small chips having high wafer yields are involved.

It was mentioned above in the discussion of leakage currents that nonlinear leakage *versus* voltage can sometimes indicate a reliability problem. A test which is sometimes used to evaluate this is known as *high current BV$_{CES}$ testing*. This test relies on crystal defects being small and nonuniformly distributed within the active area of the transistor. Transistors are measured for their collector-to-base breakdown with the testing sustained for a few seconds at a relatively high current. A peak power of approximately 15W is used to perform the screen (lower power is used to screen very low power transistors). If a localized defect is present, the majority of the current is conducted through the defect. The concentrated dissipation due to a localized defect then causes a localized melt, which normally results in a shorted transistor junction.

2.8 TRENDS IN SOLID-STATE DEVICES FOR RADAR SYSTEMS

This final section on RF power transistors discusses device trends. The history of microwave power transistors was traced at the beginning of this chapter. Early microwave power transistors were primarily targeted for use in communication systems. Those transistors were evaluated in radar applications and found to be unsuitable because of their unacceptable reliability and cost. Improvements in reliability were made by recognizing the weak areas in original designs, and by developing improved designs (e.g., geometries and processing techniques). There are three areas which have been the primary focus of those improvements:

1. Reliability improvements;
2. Performance improvements;
3. Yield improvements.

The sum of these improvements results in a cost improvement for solid-state radar systems.

Generally, the cost of microwave transistors is directly proportional to their operating frequency. Performance, in terms of the RF parameters discussed earlier, varies inversely with frequency. Because of this cost/

performance ratio *versus* frequency, solid-state radars are more competitive with their tube counterparts at lower frequencies. As with other solid-state devices, the trend in microwave power transistors will be toward improved performance and lower costs. For RF power transistors, the greatest effect will be at higher frequencies. This will result in making solid-state systems more competitive at higher frequencies as time passes. Improvements will also be made at lower frequencies, but significant cost reduction is dependent on increased device useage. Improvements will be discussed below as follows:

- Improvements in Bipolar Technology
- Other Technologies

2.8.1 Bipolar Technology

Improvements which are market-driven are always anticipated in the performance and price of solid-state devices. While a relatively large number of solid-state VHF and UHF radars have been designed and, produced, few L band systems, and even fewer S band (or higher frequency) systems have been produced. The focus of solid-state development has been in the higher frequency ranges where higher device output power, gain, and efficiency at a lower price are required in order to make solid-state radars competitive. In recent years, the focus in the L band area has been on higher power output and wider bandwidth. Device efficiency and gain, while always important parameters, have generally been adequate. At S band, gain and efficiency are too low to make solid-state widely competitive. The trend in L band bipolar technology is toward devices having output powers in the area of 300W for medium-pulse systems, and 100W for long pulse systems. In the S band range, bipolar transistors are being developed which will produce over 100W for short-pulse systems, and over 50W for medium-pulse systems. The primary focus at S band, however, is for gains above 7 dB, and efficiencies in the range of 45 percent. Table 2-3 shows forecasted device performance for UHF, L band, and S band, respectively.

2.8.2 Other Technologies

Silicon bipolar transistors have been the primary technology used in solid-state radar transmitters, but in recent years other technologies have started to challenge their preeminence. Thus far, improvements in bipolar transistor performance have kept them ahead of other technologies, but it is reasonable to expect that alternative devices may have advantages for future systems. The following paragraphs introduce the best known alternates at the time of this writing.

Bipolar transistors are current-controlled devices, which use minority carrier injection to control gain. At high temperatures, minority carriers are thermally generated. When these carriers become concentrated, they form hot spots. The hot spots have a tendency to escalate thermally, and, therefore, high power bipolar transistors have a tendency to be thermally unstable. Manufacturers of bipolar transistors have recognized this problem and have developed ballast techniques (ballast consists of local negative feedback) to counter thermal instability. The ballast method used also decreases device gain, but allows large active areas to be combined without thermal instability.

The major challenge to bipolar dominance comes from power field effect transistors (FETs). FETs are voltage controlled devices where voltage is used to control the width of the channel that conducts majority carriers. Unlike bipolar transistors, FETs have inherent thermal stability. This thermal stability permits large active areas to be combined without ballast. Unlike low power FETs, which use a horizontal channel, RF power FET's use a vertical structure. All RF power FETs currently in use are MOS FETs (sometimes called VMOS for "vertical" MOS). Various techniques are used to form the FET structure, but it is beyond the scope of this text to discuss them. FET performance, in terms of gain and efficiency, still lags behind that of bipolars. This, plus the established reliability of bipolars, has enabled bipolars to retain their position in all high-frequency radar transmitters. On the other hand, FET performance has been improving rapidly, and FETs may offer improved performance for future systems. In addition to their inherent thermal stability, the gain of FETs is easily controlled using a low-power gate bias modulator. This type of gain control can be used for pulse shaping to help control the spectrum of a radar system. FETs also have lower noise output than bipolar transistors, but this is not a problem area for radar systems. Besides their lower gain and efficiency, all FETs currently available in volume production quantities use aluminum-metal systems. Gold metalization has become the standard for use in radar systems because of improved reliability. Gold metalization is now being introduced on some FETs, but the reliability of gold metalized FETs is yet to be established (gold metalization is difficult to use with voltage-controlled devices because it causes band-gap shifts, and, therefore, significant gain changes).

As FET performance approaches that of bipolar transistors, it can be anticipated that they will be evaluated for, and used in, radar transmitters. The theoretical performance limit for both bipolar transistors and FETs is governed by their use of silicon crystals. FET performance improvements will level off as they approch those of bipolars. Other attributes of the devices will then be used to select the best technology for radar systems.

In addition to FETs, a third silicon technology has been reported as a possible source for radar transmitters: *static induction transistors* (SITs). At this time, RF power SITs have been prominent in device literature and absent from catalogs. Like FETs, static induction transistors are voltage-controlled devices, and, therefore, offer inherent thermal stability. Reports indicate that SITs also have higher input impedance than either FETs or bipolar transistors, and higher input impedance would be advantageous. Because of the limited performance information available on SITs, and the lack of reliability data, it is not likely that they will be employed in radar transmitters before the end of the 1980s.

Performance of FETs and SITs reported at the time of this writing is summarized in Tables 2-4 and 2-5, respectively.

Table 2-3 Forecast of the Performance of RF Power Bipolar Transistors Operating from UHF to S Band which is Anticipated by 1990

Frequency of Operation	Pulse	Peak P(0) Power Out (W)	Gain (dB)	Collector Efficiency (percent)
UHF	Short	1200	8.0	50
UHF	Medium	800	8.0	75
UHF	Long	500	8.0	80
L-Band	Short	1000	6.5	45
L-Band	Medium	500	7.0	50
L-Band	Long	250	6.5	55
S-Band	Short	200	5.5	30
S-Band	Medium	150	6.0	35
S-Band	Long	75	5.5	40

Table 2-4 Summary of Reported RF Power FET Performance as of 1985

Manufacturer	Part Number	Performance
Acrian	UMIL4OFT	P(O)= 40 W, Freq = 100 - 400 MHz, Gain = 13 dB, Drain Effic = 50 % Broadband, CW
	Development	P(O) = 110 W, Freq = 1100 MHz Gain = 8.6 dB, Drain Effic = 32 % Narrowband, Pulsed 10 uS @ 10 %
M/A-COM PHI	UF28100V	P(O) = 100 W, Freq = 400 MHz, Gain = 13 dB, Drain Effic = 55 %, CW, Broadband
	Development	P(O) = 12 W, Freq = 1400 MHz, Gain = 7 dB, Drain Effic = 43 % CW, Narrowband
Motorola	MRF154	P(O) = 600 W, Freq = 100 MHz Gain = 8 dB, Drain Effic = 40 %
Polycore	F1008	P(O) = 40 W, Freq = 400 MHz Gain = 13 dB, Drain Effic = 50 % Broadband, CW
	F2001	P(O) = 3 W, Freq = 1050 MHz Gain = 6.7 dB, Drain Effic = 30 %

Table 2-5 Summary of Reported RF Power SIT Performance as of 1985

Manufacturer	Part Number	Performance
GTE	Development	P(O)' = 25 W, Freq = 900 MHz, Gain = 9 dB, Drain Effic = 54 %, CW, Narrowband
	Development	P(O) = 180 W, Freq = 225 MHz, Gain = 6 dB, Drain Effic = 70 % CW, Narrowband
MSC	Development	P(O) = 47 W, Freq = 450 MHz, Gain = 6 dB, Drain Effic = 60 %, CW, Narrowband
	Development	P(O) = 20 W, Freq = 1000 MHz, Gain = 5.5 dB, Drain Effic = 40 % CW, Narrowband
Mitsubishi	Development	P(O) = 100 W, Freq = 1000 MHz, Gain = 4 dB, CW, Narrowband
Toshiba	Development	P(O) = 9.1 W, Freq = 2100 MHz, Gain = 3.6 dB, CW, Narrowband

3.1 INTRODUCTION

In recent years the advantages of the silicon bipolar transistor as a power stage in the 0.4 to 4 GHz range have been widely recognized [1-8]. The newest radar systems are calling for performance requirements that far surpass the capabilities of klystron or tube-type transmitters, but appear ideally suited to solid-state devices. Solid-state systems have demonstrable levels of performance and reliability that tube systems are simply unable to achieve. Tube power amplifiers require high-voltage power supplies, filament power, and electromechanical cavity tuning for wide bandwidths, while the solid-state replacement is a broadband device capable of operating from low-voltage power supplies.

As the building block power stage unit, a silicon bipolar transistor is the best candidate device for the frequency ranges from UHF through S band, notwithstanding the recent advances in power GaAs FET technology. Bipolar devices cost-effectively provide for system requirements of reliability, ruggedness, electrical performance, packaging, biasing, cooling, availability, and ease of maintenance. Largely due to new developments in processing technology, creative device packaging, and internal matching techniques, silicon devices are competing effectively up to S band frequencies. Bipolar devices are gaining an equal footing among competing technologies such as GaAs for frequencies above 4 GHz through the use of more shallow emitter diffusions, reductions in the collector-base time constants, submicron geometries, and more exotic photolithographic processes and etching techniques.

This chapter will describe bipolar devices for use as power stages from the UHF to S band frequency ranges. In particular, the device, its characteristics, and circuit requirements for operation as a class C amplifier will be reviewed. Eventually, solving the complex equation for the mutually exclusive parameters of high-power output levels from low impedances over very wide bandwidths becomes the dilemma of every high-power designer. An understanding of the basic concepts involved and the available tools will be presented.

3.1.1 Class of Operation

Class C operation is the preferred mode for bipolar devices because power output and efficiency are maximized therein. In class C operation, the transistor is biased below cut-off and is turned on only when the negative input swing exceeds the base-emitter diode drop. Thus, the device is turned on for less than 180 degrees of an RF cycle. The angle over which collector current flows is called the conduction angle. Efficiency is high because current flows when the device voltage is low. During the positive input swing, the base-emitter junction is reverse-biased and no current flows through the collector, but the energy stored in the collector circuit from the previous cycle is dissipated in the load. During each RF cycle, the transistor is driven through its cut-off, active, and saturation regions. Examples of the input and output waveforms are shown in Figures 3-1 and 3-2 [9,10]. Because of the complex interaction of the nonlinearities, a general analytical solution for class C operation is difficult to obtain [11].

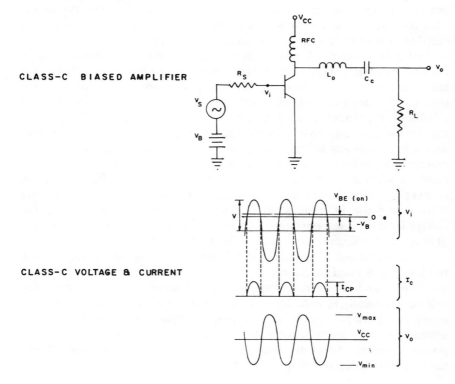

Figure 3-1

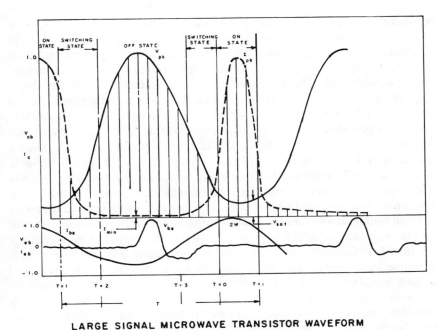

LARGE SIGNAL MICROWAVE TRANSISTOR WAVEFORM

Figure 3-2

For straightforward sinusoidal amplification, the maximum collector efficiency for a class C amplifier with conduction angle θ is given by [12]:

$$\eta_C = \frac{\text{RF Power Output}}{\text{dc Power Input}} = \frac{\theta - \sin\theta}{4 \sin(\theta/2) - 2\theta \cos(\theta/2)} \qquad (3\text{-}1)$$

Because the efficiency approaches 100 percent as the power output approaches zero, θ is made small to maximize the efficiency while still maintaining power output.

3.1.2 Device Characterization

For large-signal class C amplifiers, complete network parameters for forward and reverse power gains cannot be easily measured. Consequently, design by the S-parameter method is essentially meaningless [13]. For small-signal analysis, measurement and manipulation of the device S-parameters will yield exact behavior of the amplifier in terms of power gain and stability. In large-signal saturating power devices, great excursions in the output voltage waveform take the device through all of its available states during each RF cycle. S-parameter measurement does not handle the effects of saturation, cut-off, stability, current crowding, and base widening. The two most critical parameters of a power stage, namely, the saturated power output and the collector efficiency, cannot be accounted for by way of S-parameter description. These parameters become a function of collector load impedances. In general, the circuit designer's strategy is to determine the device performance for known source and load impedances. The designer then generates a suitable source and load network that duplicates those conditions. This is done while satisfying some minimum requirements for power output, gain, efficiency, stability, and ruggedness. The device can be characterized by optimal tuning in a fixture for minimal input power reflection and maximal output impedances, which appear at the input and output of the transistor for tuned performance that can be measured using a network analyzer. Because large power devices invariably have extremely low input and output impedances, the first-cut fixture used for characterization must be relatively close to optimal for external tuning to have any effect. A combination of creative modeling and forthright circuit design is necessary for accurate results. The general procedure is:

1. Model the terminal input and output impedances of the transistor based on its physical parameters and dc characteristics;
2. Design a first-cut impedance-matching network that transforms the modeled output impedances to 50 ohms;
3. The device must be optimally tuned to maximize power output and minimize the reflected power by using external multiple-stub tuners;
4. Measure the source and load impedances for the tuned condition;
5. Perform load-pull measurements;
6. Design the final impedance-matching circuit;
7. Optimize and verify the performance.

3.1.3 Equivalent Large-Signal Model

Characterization of a power transistor describes the ultimate dilemma in which the answer must be guessed before making the measurements that determine what the answer will be. Actually, a fairly accurate estimate for the terminal impedances can be calculated on the basis of the physical parameters of the transistor and the package. A first-cut impedance-matching network can then be designed so that an accurate characterization can be made.

As discussed in Chapter 2 on devices, a modern power transistor might actually be a complex hybrid circuit in a multilayer ceramic package. Depending on its operating frequency and power handling capability, the device may include several steps of low-pass impedance matching, output matching, and considerable parasitics. A reliable device model is an invaluable tool for evaluating an amplifier. Because amplifier performance is largely dependent on load impedances, it is important to understand how the terminal impedances of the device will change for variations in collector voltage, frequency, and RF drive level. This section not only evaluates the characteristics of the transistor chip itself, but also the impedance transforming effects of the multiple-wire bonds, MOS capacitors, and other distributed elements. The models used are simple approximations that help to predict terminal impedances. The effects caused by the harmonic components of the currents and voltages generated by the nonlinear elements of the transistor are generally neglected.

Many class C amplifiers in operation above the UHF range are operated as common-base amplifiers. The common-base configuration is often preferred because it tends to deliver higher power gain than the common-emitter configuration. Increased gain at the higher frequencies for the common-base stage is a result of regeneration caused by base-spreading resistance and common lead inductances. Because of the current requirement at the emitter, common-base input impedances are extremely low. Magnitudes of the input impedance can be on the order of one ohm or less at the chip. Figure 3-3 shows a detailed electrical model of a bipolar transistor superimposed on a sketch of its physical construction. By

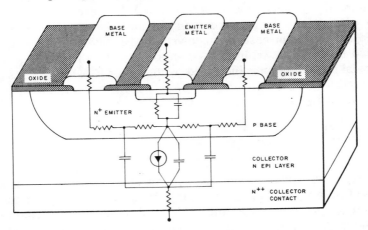

EQUIVALENT CIRCUIT SUPERIMPOSED ON TRANSISTOR
PHYSICAL STRUCTURE

Figure 3-3

assuming some of the elements to be negligible and by *a priori* knowledge of the magnitudes of the base-spreading resistance and output conductance, the models of Figures 3-4 and 3-5 are valid. Figure 3-4 shows the common-base equivalent [14], while Figure 3-5 shows a further simplification of the basic "tee" equivalent circuit [15]. Both models are accurate for evaluation above 500 MHz.

The output model for the transistor assumes a current source shunted by a conductance and the collector-base capacitance. The power transfer theorem dictates that maximum power is delivered to a load when the load is the conjugate match for the system source resistance; thus, the equivalent output resistance shunting the current generator is approximately equal to the load resistance. The output voltage swing is nearly sinusoidal, swinging about V_{CC} is limited in its negative excursion by the value of $V_{CE\ sat}$. At microwave frequencies, $V_{CE\ sat}$ can be significantly higher than the dc saturation voltage. (This occurs because the microwave current distribution is more restricted at the higher frequencies, and results in an active transistor area which is less than that at lower frequencies.) Thus, using Ohm's law:

$$P_0 = \frac{(V_{CC} - V_{CE\ sat})^2}{2\,R_L} \tag{3-2}$$

In reality, the collector voltage waveform approaches that of Figure 3-6 because class C stages tend to be driven into saturation where power output and efficiency can be optimized. For very short pulsewidth operation ($< 50\ \mu s$), where operation from a 50 Vdc collector supply voltage is not uncommon, the positive swing of the RF waveform may even exceed the collector-base breakdown voltage without incurring catastrophic failure. If a saturating waveform occurs, then the equivalent sinusoidal power output is increased by the factor K, where K is as given in Figure 3-7 [16].

The value R_L can now be approximated as

$$R_L = \frac{K(V_{CC} - V_{CE\ sat})^2}{2\,P_0} \tag{3-3}$$

This expression remains only an approximation and is valid for small ranges of R_L because $V_{CE\ sat}$ can change significantly with collector current and temperature. Thus, the value of R_L can be substituted as the intrinsic output resistance of the transistor under a narrow range of operation.

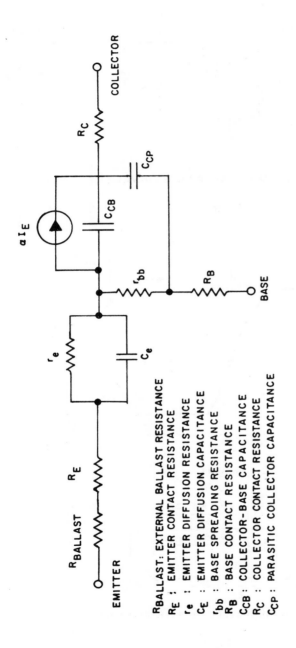

RBALLAST: EXTERNAL BALLAST RESISTANCE
R_E : EMITTER CONTACT RESISTANCE
r_e : EMITTER DIFFUSION RESISTANCE
C_E : EMITTER DIFFUSION CAPACITANCE
r_{bb} : BASE SPREADING RESISTANCE
R_B : BASE CONTACT RESISTANCE
C_{CB}: COLLECTOR-BASE CAPACITANCE
R_C : COLLECTOR CONTACT RESISTANCE
C_{CP}: PARASITIC COLLECTOR CAPACITANCE

COMMON BASE "TEE" EQUIVALENT OF A BIPOLAR TRANSISTOR

Figure 3-4

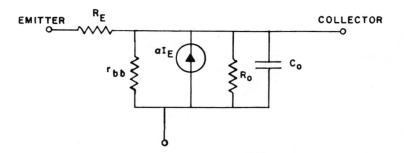

SIMPLIFIED UNILATERAL EQUIVALENT CIRCUIT

Figure 3-5

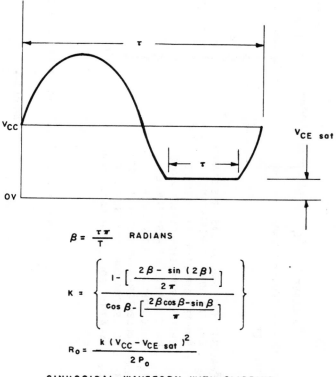

$$\beta = \frac{\tau \pi}{T} \quad \text{RADIANS}$$

$$K = \left\{ \frac{1 - \left[\dfrac{2\beta - \sin(2\beta)}{2\pi} \right]}{\cos\beta - \left[\dfrac{2\beta\cos\beta - \sin\beta}{\pi} \right]} \right\}$$

$$R_o = \frac{k (V_{CC} - V_{CE\,sat})^2}{2 P_o}$$

SINUSOIDAL WAVEFORM WITH CLIPPING

Figure 3-6

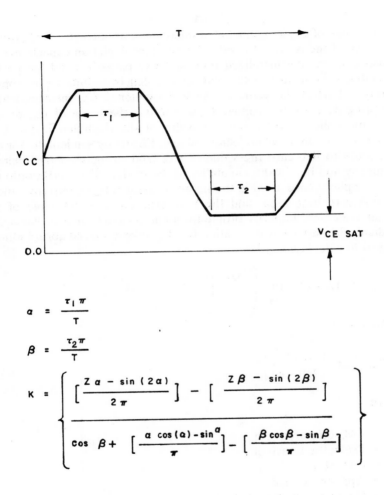

$$\alpha = \frac{\tau_1 \pi}{T}$$

$$\beta = \frac{\tau_2 \pi}{T}$$

$$K = \left\{ \frac{\left[\dfrac{Z\alpha - \sin(2\alpha)}{2\pi} \right] - \left[\dfrac{Z\beta - \sin(2\beta)}{2\pi} \right]}{\cos \beta + \left[\dfrac{\alpha \cos(\alpha) - \sin^{\alpha}}{\pi} \right] - \left[\dfrac{\beta \cos\beta - \sin\beta}{\pi} \right]} \right\}$$

SINUSOIDAL WAVEFORM WITH CLIPPING

Figure 3-7

The value of C_{cb} is significant at microwave frequencies. It is primarily the value of the reverse-biased collector-base depletion capacitance, but it also contains the metalized transistor-pad parasitics and the package capacitance parasitics. The value of C_{cb} can be estimated as approximately 1.2 times the value of the low frequency C_{ob} measured at the operating dc collector voltage [17]. Because the value of C_{ob} depends upon the collector area and the width of the depletion layer, it is a complex function of the collector voltage. Efforts by semiconductor manufacturers to decrease the value of C_{ob} tend to increase the operating frequency and bandwidth capability of the device. The Ep/Ba ratio (emitter periphery/base area) is the figure of merit relating current handling capability to base area, and the base area dictates the value of C_{ob}. Because the base impurity profile for a microwave transistor has a steep gradient, the capacitance equation for the reverse-biased abrupt junction is used to calculate C_{ob} [18]:

$$C_{ob} = A_C(4.12 \times 10^3) \left[\frac{N_C}{V + \phi} \right]^{\frac{1}{2}} \qquad (3\text{-}4)$$

or for silicon:

$$C_{ob} = A_C \left[\frac{\epsilon_r \epsilon_0 N_c}{2(V + \phi)} \right]^{\frac{1}{2}} \qquad (3\text{-}5)$$

where

A_C = Collector area (cm²)
ϵ = Dielectric constant
ϵ_0 = 8.85 pF/m
V = Applied voltage
ϕ = Contact potential of the junction

Note that increased values of V_{cb} actually decrease the value of C_{ob}.

On the input, r_{bb} represents the base-spreading resistance. Current and voltage are distributed in the base, depending upon the dimensions of the geometry and the resistivity in different parts of the base region. A transverse voltage drop can occur and r'_{bb} takes on the values of several different time constants.

Also on the input, r'_e represents the resistance measured between the active emitter region and the emitter ballast resistance.

In summary, these four parameters are: the base-spreading resistance (r'_{bb}); the emitter resistance (r'_e); the collector-base capacitance (C_{ob}); and the equivalent output resistance (Ro).

The base resistance (r'_{bb}) is a very difficult parameter to measure because it is a distributed resistance and the frequency pulse method described by Getreau may be used [19]. The collector-base capacitance (C_{ob}) is a function of reverse bias and can be measured on a capacitance bridge at the desired value of V_{CC}. This can only be accomplished on transistors without internal collector matching. Information about C_{ob} generally can be obtained from the device manufacturer. For unmatched transistors, the collector and base terminals of the device are connected to the capacitance bridge and the emitter terminal is left open. The measured value also includes the package capacitive parasitics, which can be calculated and subtracted from the measured value for a more accurate approximation of C_{ob}. The emitter resistance (r'_e) is primarily a measure of the ballast resistance added to the emitter fingers in a power transistor to equalize current distribution and reduce the generation of hot spots. The method of measurement of r'_e is also described by Getreau [19]. The emitter resistance can be determined from a curve tracer as the inverse slope of the base current as a function of collector-emitter voltage. The test set-up and typical plot are shown in Figure 3-8.

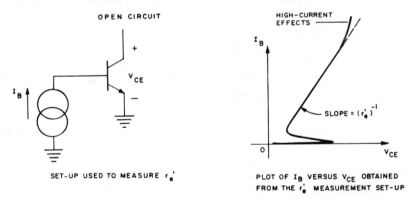

Figure 3-8

As an example, the following parameters are given for a 35W multicell power transistor operated at 1400 MHz

C_{ob} at 32 Vdc = 12pF
r'_e = 200 milli-ohms
r_{bb} = 1.0 ohm

and R_O can be calculated:

$$R_O = \frac{(32-2.0)^2}{2(35)} = 12.8\,\Omega$$

Thus, the approximate model for the intrinsic semiconductor chip is:

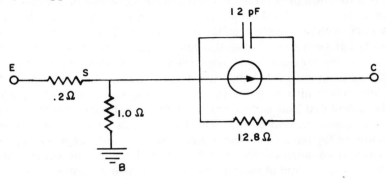

3.1.4 Package Parasitic Modeling

At microwave frequencies, package parasitics contribute significantly to the terminal input and output impedances. Distributed capacitance from printed transmission lines in a typical ceramic package, along with multiple-wire bond conductances and MOS capacitances, form a complex network. Figure 3-9 shows a view of a multiple-chip device that uses two sections of low-pass L-C impedance matching on the input and a single section of output impedance matching. For accurate representation, all the elements of the assembly should be accounted for:

- Bond-wire inductances, including self-inductance and the mutual inductances of parallel wires
- Discrete capacitance values
- Parasitic pad capacitances
- Transmission line characteristics in terms of characteristic impedance and electrical length (Z_0 and θ)

A high-power transistor may have as many as 100 individual wire bonds and a half dozen MOS capacitors, depending on the impedance matching. Because the inductance values of these wires can range from 0.1 to 2 nanohenries (nH), it is essential to model these inductances accurately at the higher frequencies. The inductive reactance of a 0.5-nH inductor in series with a common-base input resistance of one ohm yields an overall input impedance of $1.0 + j9.5$ ohms at 3 GHz. Internal, external, and mutual inductances can be figured using the following equations or graphs. The total inductance for a group of parallel wires is composed of four terms:

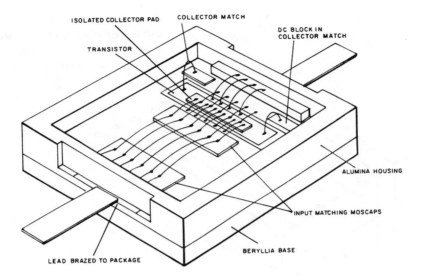

ISOLATED COLLECTOR PAD   COLLECTOR MATCH

DC BLOCK IN
COLLECTOR MATCH

TRANSISTOR

ALUMINA HOUSING

INPUT MATCHING MOSCAPS

BERYLLIA BASE

LEAD BRAZED TO PACKAGE

Figure 3-9

1. External self-inductance for a circular wire (L_{ext});
2. Internal inductance for a circular wire (L_{int});
3. Mutual inductance of coplanar segments of a wire (L_m);
4. Mutual inductance of parallel wires (L_p).

As an example, the equivalent inductance of the eight parallel wires shown in Figure 3-10 can be calculated as

$$L_{equiv} = \frac{L_{int} + L_{ext} + L_P + 2L_M}{\text{Number of Wires}} \tag{3-6}$$

L_{int} and L_{ext} are plotted in Figures 3-11 and 3-12 [20]. L_{int} is a measure of the internal inductance of a wire of circular cross section as a function of diameter and frequency, while L_{ext} is the inductance of a wire of circular cross section as a function of diameter and length:

$$L_{ext} = \frac{\mu_0 l}{2\pi} \left[\ln \left(\frac{l}{a} + \sqrt{1 + \left(\frac{l}{a}\right)^2} \right) - \sqrt{1 + \left(\frac{a}{l}\right)^2} + \frac{a}{l} \right] \tag{3-7a}$$

where l = length and a = radius,

$$L_{int} = \frac{\mu_0 l}{4\pi} \left(\frac{\delta}{a} \right) \tag{3-7b}$$

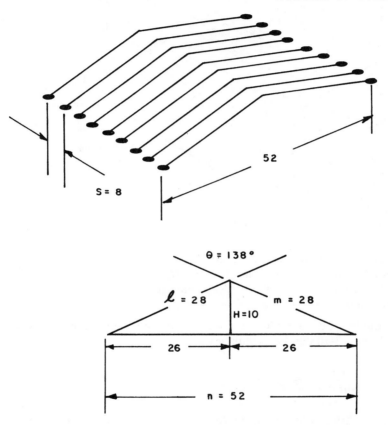

WIRE DIA = I MIL GOLD
WIRE LOOP HEIGHT = 10 MIL
WIRE SPACING = 8 MIL
NUMBER OF WIRES = 8

$$L_{EQUIV} = (L_\ell (INT) + L_\ell (EXT) + L_m (INT) + L_m (EXT) + 2L_m + L_p)/8$$

$$L_{EQUIV} = \left[(.00033)(28) + .54 + .00033(28) + .54 + 2(.078) + .0075(56)\right]/8$$

$$L_{EQUIV} = .21\, nH$$

Figure 3-10

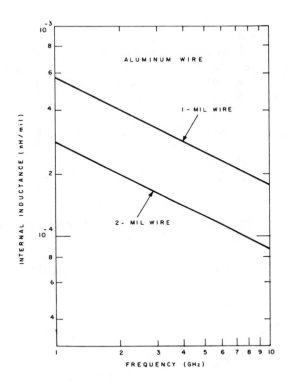

Figure 3-11

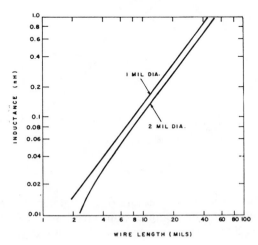

Figure 3-12

where δ is the skin depth.

L_M is a measure of the mutual inductance that occurs between two segments of the same wire offset by angle θ, while L_{Mp} is a measure of the mutual inductance that accrues due to parallel wires of length l separated by spacing S (Figure 3-13):

$$L_M = \frac{\mu_0}{2\pi} \cos \theta \left[l \tanh^{-1} \left(\frac{m}{l+n} \right) + m \tanh^{-1} \left(\frac{l}{n+m} \right) \right] \qquad (3\text{-}8a)$$

$$L_P = \frac{\mu_0}{2\pi} l \left[\ln \left(\frac{l}{S} + \sqrt{1 + \left(\frac{l}{S} \right)^2} \right) - \sqrt{1 + \left(\frac{S}{l} \right)^2} + \frac{S}{l} \right] \qquad (3\text{-}8b)$$

The total equivalent inductance at 3GHz for the bond-wire configuration in Figure 3-10 is given as

$$L_{equiv} = (L_{int} + L_{ext} + L_{int} + L_{ext} + 2\,L_M + L_P)/8 = .21\mu H \qquad (3\text{-}9)$$

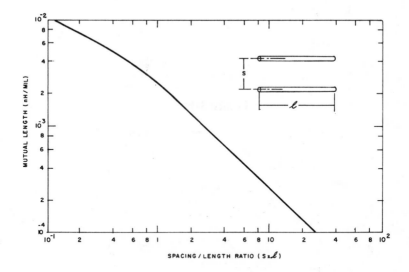

Figure 3-13

Inductor and capacitor quality factors (Q) are generally considered to be quite high for gold thermocompression wire bonds and high quality MOS capacitors used by transistor vendors. Capacitors are fabricated by depositing a metal layer over an oxide layer that has been grown on a highly doped, low resistivity silicon. Measured capacitor Q is in excess of 400 at 400 MHz for 50pF capacitors [21]. Bond-wire Q can be calculated by figuring in the resistivity of gold at a certain skin depth with the value of inductance. Table 3-1 shows how the effective resistance of various gold wire diameters (calculated for 3 skin depths) varies with frequency. For example, a 1-mil diameter gold wire, 50-mils long, exhibits an inductance of approximately 1nH. Because of skin depth variations with frequency, Q can vary from 190 at 3GHz to as low as 40 at 0.5GHz. For very low impedance operation, especially in common-base input stages, bond-wire losses can affect gain performance in improperly designed matching networks.

Table 3-1

EFFECTIVE RESISTANCE OF GOLD WIRE
CALCULATED AT 3 SKIN DEPTHS

	DC	1.4 GHz	3.0 GHz	10 GHz
SKIN DEPTH:	∞	2.0 μ in	1.4 μ m	.77 μ m
1 MIL WIRE:	1.17 Ω/in	1.68 Ω/in	2.18 Ω/in	3.61 Ω/in
1.5 MIL WIRE:	.52 Ω/in	1.0 Ω/in	1.36 Ω/in	2.35 Ω/in
2 MIL WIRE:	.3 Ω/in	.72 Ω/in	.99 Ω/in	1.73 Ω/in
3 MIL WIRE:	.13 Ω/in	.46 Ω/in	.64 Ω/in	1.13 Ω/in
1 x 10 MIL RIBBON:	.09 Ω/in	.18 Ω/in	.26 Ω/in	.45 Ω/in

Most power transistor packages are fabricated in the same way, and an understanding of the basic construction scheme aids in analysis. A multilayer hermetically sealed ceramic housing encapsulates the transistor die and internal matching. Industry standard packages are generally similar to the co-fired ceramic type shown in Figure 3-14 [22]. A beryllia base is metallized through a refractory screen in the "green" state, or the pre-fired flexible state. The base ceramic used is beryllia (BeO) because of its excellent thermal conductivity. The side walls, end walls, and microstrip feeds are built on alumina ceramic (Al_2O_3). The beryllia and alumina are co-fired for adhesion at temperatures in excess of 1500 degrees Fahrenheit, hence the need for refractory metallization. Thin-film deposited metals, such as a chrome-gold system, are not compatible with these temperatures. Refractory metals such as molybdenum, tungsten, or manganese will sustain these temperatures, and they are compatible with gold-plating techniques. Because of the furnace firing sequence required, the beryllia and alumina must be of degraded purity as compared with the ceramic quality used in thin-film techniques. In general, metallic losses are higher due to the more "resistive" metals used. Also, the surface finish in the ceramic is poor and so the loss tangent is higher.

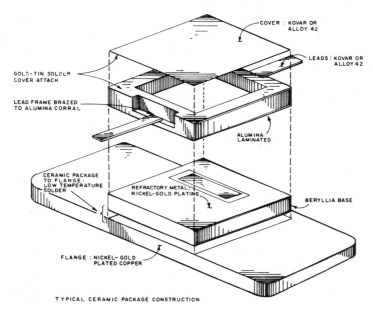

Figure 3-14

The material effects all become significant at higher frequencies and lower impedances. Table 3-2 lists the properties of several ceramics and metals encountered in package design, and Figure 3-15 depicts a physical layout for a multicell device. Figure 3-16 illustrates the equivalent model, which has been used to predict the terminal impedances based on physical parameters, while Figure 3-17 shows the measured impedances for this particular device.

Table 3-2

	RESISTIVITY ($\mu\Omega-cm$)	TCE THERMAL COEFFICIENT OF EXPANSION min/in/°F
ALLOY 42*	—	5.0
ALUMINUM	2.6	13.1
CHROMIUM	13	3.4
COPPER	1.7	9.2
GOLD	2.2	7.9
KOVAR	45 – 85	2.7
MANGANESE	185	12.0
MOLYBDENUM	5.2	2.8
NICKEL	6.8	7.4
PALLADIUM	10.8	6.6
PLATINUM	9.8	4.9
SILVER	1.6	10.9
TIN	11.5	13.0
TUNGSTEN	5.5	2.4

SOME COMMONLY FOUND METALS* IN PACKAGING

	THERMAL CONDUCTIVITY (BTU/hr.ft.° F)	TCE min/in/°F	DIELECTRIC CONSTANT (er)
ALUMINA (99%)	14.5	3.7	9.6 - 10
ALUMINA (95%)	12.1	3.7	8.9
ALUMINA (85%)	8.5	3.7	8.2
BERYLLIA (99%)	95.2	5.3	6.8
SILICON	49.0	2.4	——
SAPPHIRE	15.9	3.8	9.4 – 11.6
GALLIUM ARSENDE	25.6	3.4	12.5

INSULATORS FOUND IN PACKAGING

*[23]

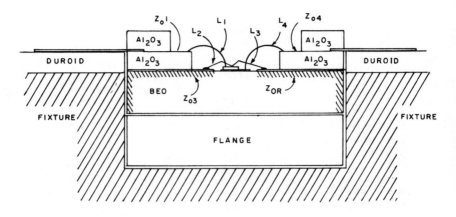

CROSS SECTION OF AN UNMATCHED HERMITIC
S-BAND BIPOLAR TRANSISTOR

Figure 3-15

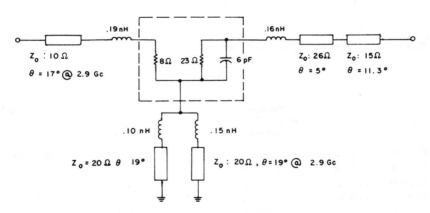

EQUIVALENT ELECTRICAL MODEL FOR PACKAGED DEVICE

Figure 3-16

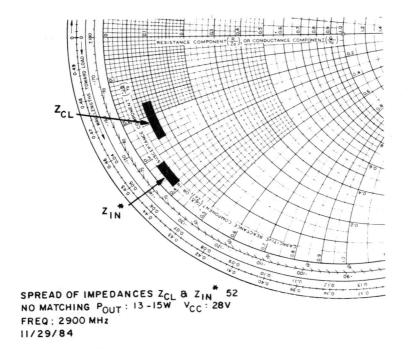

SPREAD OF IMPEDANCES Z_{CL} & Z_{IN}^{*} 52
NO MATCHING P_{OUT} : 13 -15W V_{CC} : 28V
FREQ : 2900 MHz
11/29/84

Figure 3-17

3.1.5 RF Test Set-Up

A typical test set-up for high-power device characterization is shown in Figure 3-18. A low-level sweep generator can be used to power a solid-state power amplifier or traveling wave tube (TWT). A frequency counter is used to monitor both the operating frequency and the pulsewidth/pulse repetition frequency if the device is to be operated in the pulsed mode. Power measurements can be made on an average basis, where the peak power is calculated by dividing the measured average power by the duty cycle. A peak-reading power meter would reduce the aggravation. Input power and reflected input power are monitored using a high-directivity dual directional coupler.

The circulator following the TWT not only provides output protection for the TWT, but also maintains a constant reference source impedance. This is important when the transistor conjugate input impedances must be measured by output power and a video display of the output pulse coupler. Strict attention should be focused on the equipment train following the device under test (DUT). It is good practice to maintain at least

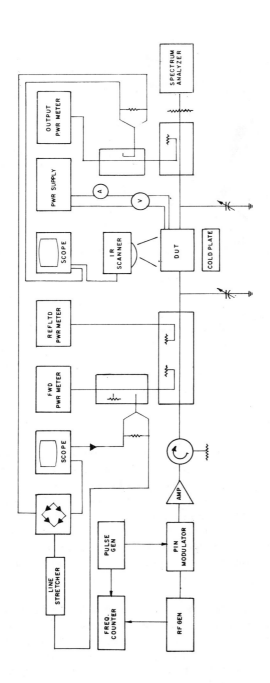

R F   P O W E R   M E A S U R E M E N T   T E S T   S E T

Figure 3-18

1.1:1 load VSWR for accurate results. A sample of the input power is mixed with a sample of the ouput power in a diode bridge such that the dc output level can be used to measure changes in insertion phase. A line stretcher is used for calibration. Peak junction temperature is a crucial measurement for determining the device's power handling capabilities. This temperature can be measured on a transient basis using an infrared scanner. The dc voltage and current flowing to the device are measured such that collector efficiency can be calculated. A spectrum analyzer is necessary to determine the harmonic content of the output and the level of spurious frequencies, which might otherwise go undetected. Finally, a cold plate or hot plate that maintains an even background temperature rounds out the set-up.

3.1.6 Measuring the Tuned Characteristics

As an example, let us say we have a hypothetical device to be measured, which is claimed by the manufacturer to be a 55W device, operated in CW at 950MHz with 7dB gain and 45 percent efficiency. Characterization of this device requires, ahead of time, an impedance-matching network, which enables the device to deliver output power to within 2dB of the device ratings, collector efficiency to within 15 percentage points of the rated efficiency, and an input match that is at least a 2.5:1 VSWR. Otherwise, circuit losses that can accummulate with relatively long transmission line lengths between low impedance at the device and the external tuners will significantly mask the true capabilities of the transistor. Figure 3-19 indicates the collector load impedance required for this 55W transistor. Note that if the collector load impedance of the first-cut circuit were situated at point Y, the tuner would not be effective in delivering the required load impedance, i.e., $2.5 - j6.5$ ohms, for maximum power output. The loci around point Y represent the effective range of the external tuners. In this case, a 4:1 VSWR at 950MHz represents a practical limit.

For an amplifier tuned at each operating point, the *dynamic range* of the device is evaluated. For class C amplifiers, there is a very small range of input power for which the output exhibits *linear* characteristics, i.e., there is a linear range in the transfer function. A nominal operating point is often chosen just at the knee of saturation, as in Figure 3-20. Highest efficiency is generally attained at this point. The collector voltage for half CW-type parts should be no more than half the value of BV_{CES}, but can be significantly higher for short pulse, low duty cycle operation. RF drive can then be varied to determine the power transfer characteristics of the device. Input power is varied from very low levels to the point where

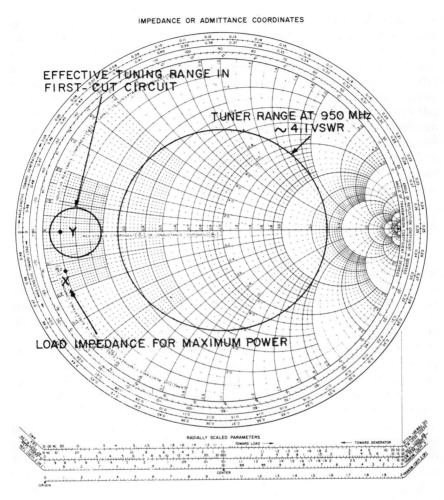

IMPEDANCE OR ADMITTANCE COORDINATES

EFFECTIVE TUNING RANGE IN FIRST-CUT CIRCUIT

TUNER RANGE AT 950 MHz ~ 4:1 VSWR

LOAD IMPEDANCE FOR MAXIMUM POWER

Figure 3-19

the output power reaches saturation. Adjustable tuning using the external variable stub tuners will be required for each incremental change in drive and output power. Because the output impedance of the device undergoes radical changes with varying power levels, an output match must be carefully maintained as the drive is increased. It is good practice to operate the device at low pulse and low duty (10µs, 10%) until the impedance-matching networks on the input and output are near optimum. This reduces the amount of dissipation within the device and may prevent a transistor from premature burn-out due to excessive load VSWR. As drive is increased, collector current (I_C), collector voltage (V_{CC}), RF power input and output (P_{in} and P_{out}), and the cell-to-cell junction temperatures must be monitored accurately in order to determine gain, collector efficiency, dissipated power, thermal resistance, and return loss. Operating points for devices will vary depending on the ratio of peak-to-average output power, frequency, drive, and thermal requirements. Some rules of thumb are:

1. Operating at 0.1 to 0.6 dB compressed gain yields the highest efficiency;
2. For increases in collector voltage (at fixed drive), the output power and gain will increase by

$$20 \log \left[\frac{V_{CC} - \Delta V}{V_{CC}} \right] \qquad (3\text{-}10)$$

or approximately .25dB/V;
3. Efficiency usually decreases as the collector voltage is reduced.

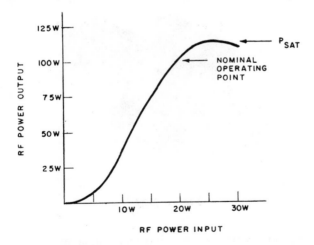

Figure 3-20

Source impedances that provide the best return loss and load impedances that provide the maximum power output must be measured. Some common nomenclature is shown in Figure 3-21, and Figure 3-22 shows a typical pull apart test fixture used for device characterization. The input and output ports of the device are designated as Z_{in} and Z_{out}, respectively, and the impedances presented to the transistor input and output are $Z_{in}*$

and Z_{cl}, respectively; for example, a device with input impedance of 4.0 to $j8.5$ ohms is conjugately matched when presented with Z_{in}^* of $4.0 - j8.5$ ohms.

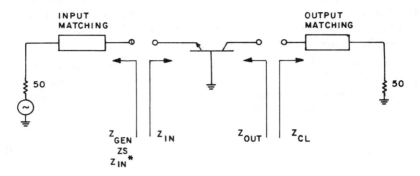

Figure 3-21

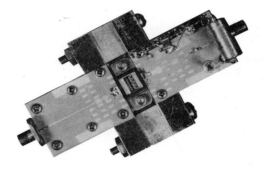

Figure 3-22

The accuracy with which source and load impedances are measured is critical. A measurement error of only 4 electrical degrees at the 2 ohm level can result in an error that is equivalent to a 2:1 VSWR. This is illustrated in Figure 3-23. Higher device terminal impedances, as a result of internal impedance matching, reduce this source of measurement error and lead to more repeatable matching circuit impedances, and hence more consistent amplifier performance.

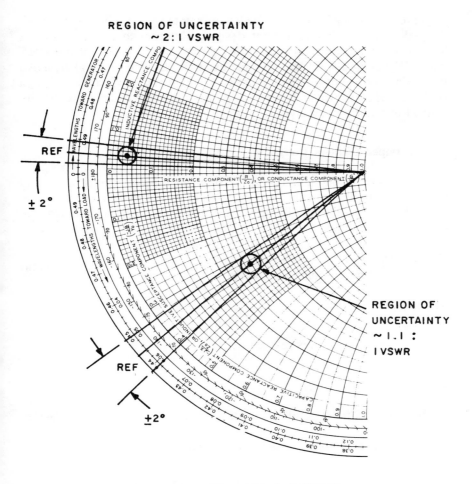

REGION OF UNCERTAINTY
~ 2:1 VSWR

REF

± 2°

REGION OF
UNCERTAINTY
~ 1.1 :
1 VSWR

REF

±2°

EFFECTS OF MEASUREMENT ERROR
AT DIFFERENT IMPEDANCE LEVELS

Figure 3-23

3.2 IMPEDANCE MATCHING

As outlined in the previous section, the input and output impedances
of a microwave power transistor can be very low and quite reactive.
Impedances become increasingly lower as the power output capability of
the device increases. These low impedances must be transformed to 50-
ohm levels, sometimes for very wide bandwidths. This section is intended

to serve as a design aid for impedance matching and will address both lumped- and distributed-element designs. External to the transistor, a lumped element design is useful through the upper ranges of the UHF band. Higher frequency designs, for example, L band and higher, make more efficient use of distributed networks. Monolithic microwave design is the exception, where lumped elements are used almost exclusively instead of a distributed layout.

3.2.1 Lumped Element Design

For a lumped element design, a practical application of circuit Q (circuit selectivity) is its use in transforming series networks into parallel equivalents or *vice versa*. The general definition of circuit Q at resonance is

$$Q = \frac{2\pi \times \text{Energy stored per cycle}}{\text{Energy dissipated per cycle}} \tag{3-11a}$$

which is also given by

$$Q_S = \frac{\omega_0 L}{R} \text{ for a series resonant circuit} \tag{3-11b}$$

and

$$Q_P = \omega_0 RC \text{ for a parallel resonant circuit} \tag{3-11c}$$

The following transformations are illustrated:

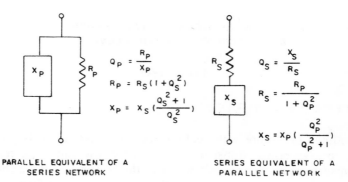

PARALLEL EQUIVALENT OF A
SERIES NETWORK

SERIES EQUIVALENT OF A
PARALLEL NETWORK

The first step in designing an input matching network is to resonate the inductive component of the input impedance with a shunt capacitance. The inductive nature of the input impedance arises because of the series bond-wire inductances and lead inductance internal to the packaged transistor. Below the UHF range, the value of inductive reactance may actually be very low, and some external series inductance may have to be added in order to build a first-section network. Alternate series L and shunt C sections are added in low-pass configuration to bring the real part of the general ladder network impedance to 50 ohms, as shown in Figure 3-24. For an example using the lumped element approach, the terminal input impedance of a device will be transformed from $1.0 + j2.0$ ohms to 50 ohms at a center frequency of 500 MHz. Typical design considerations are minimum circuit losses, broadband operation, repeatable performance, and ease of manufacture. Practical designs can be built easily in the form of multiple cascaded stages of low-pass filter networks to provide impedance matching to the 50-ohm level. To start, the input impedance should be transformed upwards, incrementally by individual networks, with Q as low as possible for bandwidth preservation. A stage Q no greater than 2 is often sufficient. The input impedance of the device is given as

$$X_S = 1 + j2\Omega \text{ and } Q_S = 2 \qquad\qquad (3\text{-}12)$$

Transforming the network to the equivalent parallel network yields

$$R_{P1} = R_S[I + Q^2] = 5\Omega$$

$$X_{P1} = X_S\left[\frac{Q^2 + I}{Q^2}\right] = +j2.5\Omega \qquad\qquad (3\text{-}13)$$

Because $Q = 2$ and $R_P = 2.5$, the value of X_C used to resonate the input inductive reactance is calculated:

$$X_C = -j2.5\Omega \qquad\qquad (3\text{-}14)$$

Now, add a second L-C stage, also with $Q_2 = 2$

$$R_{P1} = R_{S2} = 5.0$$
$$X_S = Q_2 R_{S2}$$
$$Q_2 = 2$$
$$X_{S2} = 2(5)$$
$$= +j10\Omega$$

(3-15)

Convert to the parallel equivalent:

$$R_{P2} = R_{S2}(1+Q^2) = 25\Omega$$

$$X_{P2} = X_{S2}\left(\frac{Q_2+1}{Q^2}\right) = j12.5\Omega$$

(3-16)

In order to resonate, $X_{P2} = -j12.5$ ohms. Use the last section to transform from $R_P = 25$ ohms to 50 ohms and determine the required Q for this section. Set $R_{P3} = R_{P2} = 25$ ohms:

$$Q_3 = \sqrt{\frac{R_{P3}}{R_{S3}} - 1} = 1$$

(3-17)

Therefore, the last section is

$$\left.\begin{array}{r} X_{S3} \\ R_P \\ X_P \end{array}\right\} = Q R_{S3} = +j25\Omega,$$

therefore $X = -j50$

(3-18)

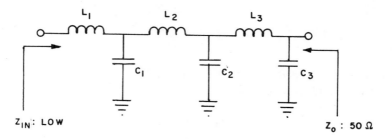

Z_{IN}: LOW Z_o : 50 Ω

LOW PASS MULTIPLE POLE IMPEDANCE
MATCHING LADDER NETWORK

Figure 3-24

The final network is represented in Figure 3-25a and calculated values of inductance and capacitance are shown in Figure 3-25b. A plot of the VSWR for the input matching network is shown in Figure 3-26. Note the broadband nature of the response that has resulted from using multiple low-Q networks for the transformation. In general, designing for wider bandwidth in a narrowband requirement will yield an easily repeatable, well-matched design for production purposes. Aberrations such as component tolerances and component placement on microwave printed circuit boards become less critical.

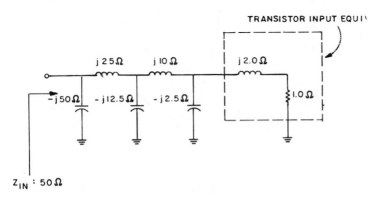

REACTANCE VALUES FOR INPUT MATCHING

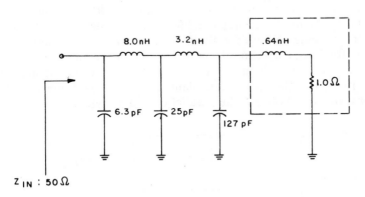

COMPONENT VALUES FOR INPUT
MATCHING @ 500 MHz

Figure 3-25

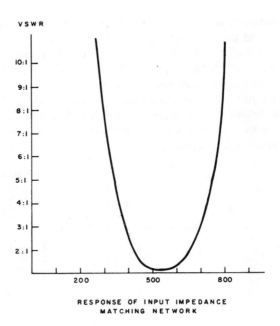

VSWR

10:1

9:1

8:1

7:1

6:1

5:1

4:1

3:1

2:1

200   500   800

RESPONSE OF INPUT IMPEDANCE
MATCHING NETWORK

Figure 3-26

The design of the output network is somewhat different. Assume once again that the center frequency is 500MHz and the device has no internal collector matching. At sufficiently low frequencies, the value of C_{ob} will dominate the imaginary component of the impedance on a device without internal collector matching. Assume that the device output impedance has been given as the series equivalent, $Z_s = 3.8 - j4.2$ ohms, and then transform this to 50 ohms. The parallel equivalent yields

$$Q_S = \frac{4.2}{3.8} = 1.1$$

$$R_P = 8.4\Omega$$
$$X_{P1} = -j7.7\Omega \tag{3-19}$$

Resonating the value of $-j7.7$ ohms with a shunt-mounted inductive reactance yields an equivalent parallel network resistance of 8.4 ohms at center frequency:

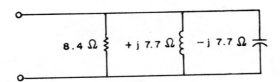

Add one low-pass L-C section with $Q = 1.6$ and the network shown in Figure 3-27 will result. The lumped-element values for inductors and capacitors are also shown. These are simply impedance-matching networks without bias components or stability enhancements. A polar plot of the output response is given in Figure 3-28.

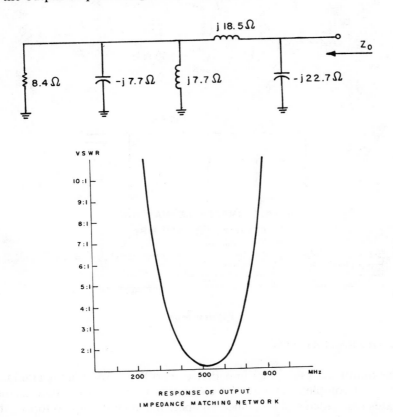

RESPONSE OF OUTPUT
IMPEDANCE MATCHING NETWORK

Figure 3-27

84

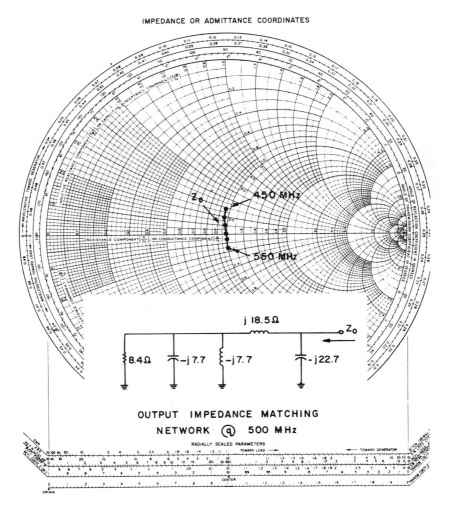

Figure 3-28

3.2.2 Graphical Analysis

The Smith chart is a polar impedance chart that is used in the graphical analysis of complex networks. It is a visual representation that accommodates resistance and reactance components in either of two forms: the admittance plane or the impedance plane. The impedance chart can be normalized to any resistance and can denote any impedance of the form $R + jX$ ohms, while the admittance chart is the inverse, representing admittances of the form $Y = G + jB$ mhos. A handy design tool is a Smith chart with an admittance plot overlaid on the impedance plot. Figures 3-29 and 3-30 illustrate the direction of movement for reactive components on impedance and admittance charts, respectively. Figure 3-31 is a graphical plot of the input impedance-matching design previously used. The input impedance of the transistor was given as $1.0 + j2.0$ ohms and is so indicated. Because the chart is normalized to 50 ohms, the impedance is plotted as $.02 + j.04$ ohms. Each stop on the chart represents the absolute impedance at points within the network. Ladder-type

IMPEDANCE OR ADMITTANCE COORDINATES

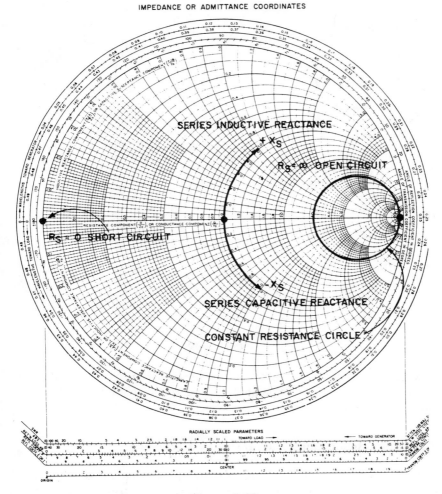

Figure 3-29

86

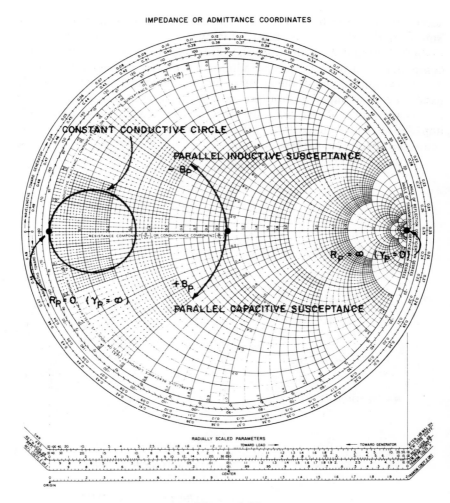

Figure 3-30

impedance-matching networks can be designed using only the Smith chart, which can be manipulated to understand circuit response as a function of frequency.

3.2.3 Microstrip Elements and Distributed Matching

The previous examples of impedance transformation serve to demonstrate how lumped-element capacitors and inductors can be used as

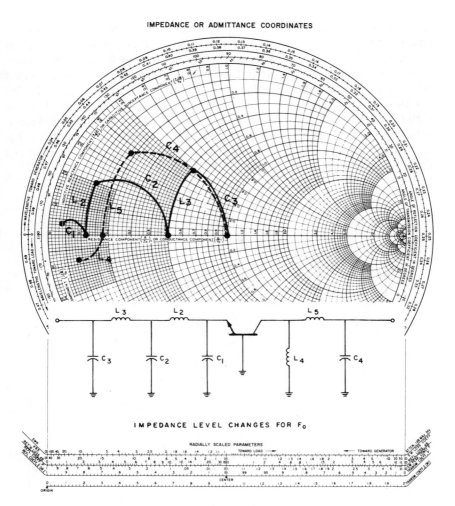

Figure 3-31

impedance-matching components. In reality, discrete wound inductor elements are seldom used above UHF frequencies for anything other than RF chokes in bias networks, and we would be hard-pressed to measure anything short of an inductance from even a high quality 200pF chip capacitor at 3.5GHz. Inductive and capacitive properties can be exploited quite easily at the higher frequencies using various configurations of quasi-TEM mode propagation in the microstrip transmission line format. This is the impedance-matching medium of choice for a majority of applications. Transmission line characteristics that are defined for characteristic impedance and electrical length can be designed by using microstrip elements of the proper physical dimensions. Figure 3-32 illustrates the layout of a microstrip transmission line and the equations that relate the characteristic impedance to the physical geometry [24]. Tables 3-3 through 3-7 relate the various W/H ratios to Z_0 and the effective dielectric constants (ϵ_{reff}) for some commonly used microwave dielectric substrates [25,26]. High quality substrates with dielectric constants ranging from 2.2 to 11 are commonly available. The properties of some of these materials are given in Table 3-8 [27,28].

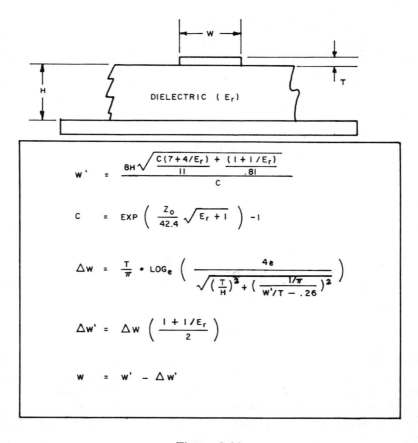

Figure 3-32

Table 3-3

```
WIDTH/HEIGHT RATIOS CALCULATED FOR THE MICROSTRIP FORMAT
DIELECTRIC NAME IS   XXXXX
DIELECTRIC CONSTANT IS    2.33
*********************************************************************
               WIDTH/HEIGHT        REL PHASE VEL.       ZO

                22.01482             .6734029            10
                17.94605             .6761468            12
                15.05169             .6787858            14
                12.89079             .6813302            16
                11.2184              .6837824            18
                 9.887531            .6861424            20
                 8.804669            .6884112            22
                 7.907464            .6905908            24
                 7.152762            .6926843            26
                 6.509748            .6946953            28
                 5.955831            .6966279            30
                 5.474082            .698486             32
                 5.051564            .7002741            34
                 4.678227            .7019959            36
                 4.346144            .7036555            38
                 4.048987            .7052565            40
                 3.781643            .7068026            42
                 3.539936            .7082968            44
                 3.320429            .7097426            46
                 3.120263            .7111426            48
                 2.937047            .7124994            50
                 2.768764            .7138156            52
                 2.613703            .7150936            54
                 2.470404            .7163354            56
                 2.337614            .7175427            58
                 2.21425             .7187175            60
                 2.099378            .7198614            62
                 1.992181            .7209757            64
                 1.891947            .7220618            66
                 1.798049            .7231207            68
                 1.709936            .7241536            70
                 1.627121            .7251611            72
                 1.549169            .7261442            74
                 1.475696            .7271035            76
                 1.406357            .7280394            78
                 1.340843            .7289524            80
                 1.278877            .7298428            82
                 1.220207            .7307108            84
                 1.164608            .7315565            86
                 1.111874            .73238              88
                 1.061817            .7331815            90
                 1.014269            .7339608            92
                  .969072            .734719             94
                  .9260842           .7354535            96
                  .8851746           .736167             98
                  .8462219           .7368589           100
                  .8091146           .7375293           102
                  .77375             .738179            104
                  .7400321           .738808            106
                  .7078724           .7394172           108
                  .6771879           .7400071           110
```

Table 3-4

```
WIDTH/HEIGHT RATIOS CALCULATED FOR THE MICROSTRIP FORMAT
DIELECTRIC NAME IS   XXXXX
DIELECTRIC CONSTANT IS    2.55                       .
***************************************************************************
               WIDTH/HEIGHT              REL PHASE VEL.        ZO
```

WIDTH/HEIGHT	REL PHASE VEL.	ZO
20.98739	.6456088	10
17.1003	.6485208	12
14.33566	.6513205	14
12.272	.6540165	16
10.67515	.65661	18
9.404634	.6591016	20
8.371043	.6614927	22
7.514793	.6637865	24
6.794636	.6659867	26
6.181118	.6680977	28
5.65266	.6701245	30
5.193079	.6720715	32
4.790017	.6739438	34
4.433876	.6757457	36
4.117087	.6774816	38
3.833607	.6791556	40
3.578555	.6807717	42
3.347949	.6823333	44
3.13851	.6838437	46
2.947513	.6853062	48
2.772676	.6867235	50
2.61208	.6880982	52
2.464095	.689433	54
2.32733	.6907297	56
2.200592	.6919905	58
2.082853	.6932172	60
1.973222	.6944114	62
1.870923	.6955745	64
1.775277	.6967079	66
1.68569	.6978125	68
1.601637	.6988895	70
1.522654	.6999395	72
1.448329	.7009634	74
1.378295	.7019615	76
1.312224	.7029344	78
1.249822	.7038823	80
1.190824	.7048056	82
1.134991	.7057041	84
1.082108	.7065781	86
1.031977	.7074277	88
.9844209	.7082528	90
.9392756	.7090533	92
.8963928	.7098296	94
.8556349	.7105818	96
.8168765	.7113102	98
.7800011	.7120152	100
.7449017	.7126973	102
.7114793	.7133573	104
.6796415	.7139958	106
.6493026	.7146136	108
.620383	.7152116	110

Table 3-5

```
WIDTH/HEIGHT RATIOS CALCULATED FOR THE MICROSTRIP FORMAT
DIELECTRIC NAME IS   BERYLLIA
DIELECTRIC CONSTANT IS    6.8
***************************************************************
          WIDTH/HEIGHT          REL PHASE VEL.        ZO

          12.31418              .4088465              10
          9.957142             .4126201              12
          8.285403             .4161688              14
          7.04091              .419503               16
          6.080352             .4226369              18
          5.317792             .4255871              20
          4.698599             .42837                22
          4.186404             .4310011              24
          3.756084             .4334947              26
          3.389744             .4358637              28
          3.074306             .4381196              30
          2.800005             .4402729              32
          2.559408             .4423326              34
          2.346769             .4443064              36
          2.157578             .4462015              38
          1.988253             .4480235              40
          1.835912             .4497776              42
          1.698214             .4514677              44
          1.57324              .4530972              46
          1.459399             .4546685              48
          1.355364             .4561837              50
          1.260019             .4576439              52
          1.172415             .4590497              54
          1.091744             .4604015              56
          1.017309             .4616992              58
          .9485086             .4629427              60
          .8848178             .4641323              62
          .8257763             .4652684              64
          .7709793             .4663521              66
          .7200679             .4673851              68
          .6727222             .4683694              70
          .6286561             .469308               72
          .5876135             .4702035              74
          .549362              .471059               76
          .5136923             .4718776              78
          .4804136             .4726618              80
          .4493525             .4734143              82
          .4203499             .4741372              84
          .3932608             .4748324              86
          .3679512             .4755018              88
          .3442981             .4761468              90
          .3221885             .4767687              92
          .3015174             .4773686              94
          .2821878             .4779476              96
          .26411               .4785064              98
          .2472006             .4790463              100
          .2313823             .4795675              102
          .2165832             .480071               104
          .2027363             .4805573              106
          .1897793             .4810271              108
          .1776543             .4814809              110
```

Table 3-6

```
WIDTH/HEIGHT RATIOS CALCULATED FOR THE MICROSTRIP FORMAT
DIELECTRIC NAME IS   ALUMINA
DIELECTRIC CONSTANT IS    9.7
*****************************************************************
```

WIDTH/HEIGHT	REL PHASE VEL.	ZO
10.07597	.34687	10
8.113936	.3506284	12
6.724262	.354121	14
5.69098	.3573704	16
4.894203	.3604012	18
4.262104	.3632366	20
3.749067	.3658982	22
3.324778	.3684053	24
2.968334	.3707746	26
2.664877	.3730207	28
2.403577	.375156	30
2.176367	.3771912	32
1.977125	.3791352	34
1.801123	.380995	36
1.644663	.3827766	38
1.504806	.3844844	40
1.379192	.3861216	42
1.265901	.3876901	44
1.163356	.3891911	46
1.070251	.3906248	48
.9854866	.3919913	50
.9081369	.3932903	52
.8374116	.3945224	54
.7726301	.3956888	56
.7132052	.3967919	58
.658623	.3978348	60
.6084336	.3988214	62
.5622392	.3997561	64
.519687	.4006434	66
.4804622	.4014871	68
.4442825	.402291	70
.4108943	.4030583	72
.3800683	.4037919	74
.3515972	.4044941	76
.325292	.4051668	78
.3009816	.4058118	80
.2785088	.4064306	82
.2577306	.4070245	84
.2385157	.4075947	86
.2207439	.4081422	88
.2043045	.408668	90
.189096	.4091729	92
.1750248	.4096579	94
.1620049	.4101236	96
.1499568	.4105709	98
.1388073	.4110004	100
.1284888	.4114128	102
.118939	.4118089	104
.1101003	.4121892	106
.1019195	.4125545	108
9.434732E-02	.4129052	110

Table 3-7

```
WIDTH/HEIGHT RATIOS CALCULATED FOR THE MICROSTRIP FORMAT
DIELECTRIC NAME IS   XXXXX
DIELECTRIC CONSTANT IS    10.5
*****************************************************************
          WIDTH/HEIGHT        REL PHASE VEL.        ZO

          9.628656            .3344112             10
          7.745681            .3381491             12
          6.412431            .3416136             14
          5.421361            .3448303             16
          4.657284            .3478255             18
          4.051196            .3506243             20
          3.559304            .3532492             22
          3.152509            .3557199             24
          2.810761            .3580537             26
          2.519816            .3602651             28
          2.2693              .3623668             30
          2.05149             .3643691             32
          1.860524            .3662807             34
          1.691885            .3681084             36
          1.54203             .3698575             38
          1.408151            .3715318             40
          1.28799             .3731338             42
          1.179707            .3746649             44
          1.081794            .3761256             46
          .9929937            .3775157             48
          .9122524            .3788352             50
          .8386785            .3800845             52
          .7715091            .381265              54
          .7100879            .382379              56
          .6538455            .3834302             58
          .6022844            .3844228             60
          .554967             .3853617             62
          .5115065            .3862513             64
          .4715588            .3870963             66
          .4348172            .3879003             68
          .4010058            .3886668             70
          .3698767            .3893987             72
          .3412056            .3900984             74
          .3147898            .390768              76
          .2904449            .3914092             78
          .2680029            .3920235             80
          .247311             .3926124             82
          .2282292            .3931769             84
          .2106295            .3937183             86
          .1943947            .3942374             88
          .1794174            .3947351             90
          .1655987            .3952126             92
          .1528482            .3956703             94
          .1410824            .3961094             96
          .1302246            .3965303             98
          .1202042            .396934             100
          .1109564            .3973208            102
          .1024211            .397692             104
          9.454327E-02        .3980478            106
          8.727216E-02        .3983891            108
          8.056075E-02        .3987165            110
```

Table 3-8

COMMONLY USED MICROWAVE SUBSTRATE MATERIALS		
MATERIAL	DIELECTRIC CONSTANT (ϵ_r)	DISSIPATION FACTOR (10^{-4}) (TAN δ) @ 10 Gc
WOVEN TEFLON FIBERGLASS	2.55	15 - 20
MICROFIBER TEFLON GLASS	2.40	4 - 8
POLYOLEFIN	2.32	3 - 5
POLYPHENELENE OXIDE	2.55	16
CROSS—LINK POLYSTYRENE	2.53	2.5 - 6.6
QUARTZ	3.8	1
BERYLLIA (99 %)	6.8	1
GALLIUM ARSENIDE	13	6
SILICON	12	10 -100
ALLUMINA (99.5%)	10	1 - 2
(96 %)	9	6
(85%)	8	15
CERAMIC-PTFE	10.5	20
EPOXY GLASS	4.5 - 4.9	180

A microstrip transmission line can have properties identical to those of a discrete element inductor or capacitor for a single frequency, or over a very narrow bandwidth. The same microstrip configuration may have wildly different characteristics at other frequencies, in particular for frequencies on the order of half an octave away. An understanding of these properties begins with knowledge of the general form of the transmission line equation. For a transmission line of characteristic impedance Z_0 and electrical length θ, terminated by load impedance Z_L, the impedance looking into that network is given as

$$Z_{in} = Z_0 \left[\frac{Z_L + jZ_0 \tan\theta}{Z_0 + jZ_L \tan\theta} \right] \qquad (3\text{-}20)$$

There are some special cases which can be quite useful to the circuit designer. These cases are:

1. $Z_L = 0$
2. $Z_L = \infty$
3. $\theta = 90$ degrees
4. very small θ (<45 degrees)

When $Z_L = 0$, the transmission line equation reduces to:

$$Z_{in} = +jZ_0 \tan\theta \qquad (3\text{-}21)$$

The tangent function is graphed in Figure 3-33. Note that the tangent function is positive for values from 0 to 90 degrees and from 180 to 270 degrees. Thus, for a length of transmission line terminated in $Z_L = 0$ ohms and for $0 < \theta < 90$ degrees, or $180 < \theta < 270$ degrees, the input impedance appears to be inductive. For example, at 1400 MHz, a shunt-mounted 10-nH inductor and a section of short-circuited transmission line 41 degrees long with characteristic impedance $Z_0 = 100$ ohms are interchangeable components:

$$Z_L = +j2\pi f_L = +j88\Omega \qquad (3\text{-}22a)$$

and for $Z_0 = 100\Omega$ and $\theta = 41°$

$$Z_{in} = +j100 \tan 41° = +j87.8\Omega \qquad (3\text{-}22b)$$

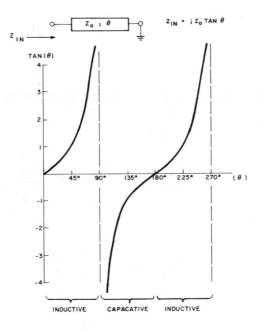

TANGENT FUNCTION

Figure 3-33

Likewise, the general form of the transmission line equation can be manipulated by L'Hospital's rule for the case where the terminating impedance is infinity, i.e., an open circuit. The input impedance is given by

$$Z_{in} = \frac{-jZ_0}{\tan\theta} \qquad (3\text{-}23)$$

such that a capacitive reactance can be modeled by a section of open-circuited transmission line where $0 < \theta < 90$, or $180 < 0 < 270$ degrees. A 10-pF capacitor at 1400MHz has the same capacitive reactance as an open-circuited stub 53 degrees long with $Z_0 = 15$ ohms:

$$Z_c = j/2\pi f_c = -j7.5\Omega \qquad (3\text{-}24a)$$

and for $Z_0 = 15$ and $\theta = 53°$

$$Z_{in} = \frac{-j15}{\text{Tan}53°} = -j7.5\Omega \qquad (3\text{-}24b)$$

Note that for small θ values, the distributed line equivalents are nearly identical to the reactance response of the lumped elements. This is illustrated in Figure 3-34.

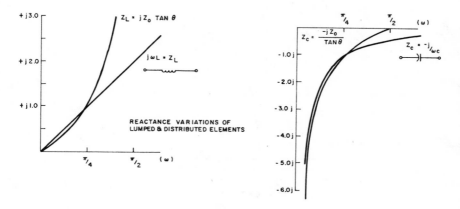

Figure 3-34

An additional circuit configuration that occurs for line lengths of exactly 90 or 180 degrees is of great use to the circuit designer. A short-circuited stub exactly 90 degrees long appears to be an open circuit. This is a useful element for a bias injection stub where no perturbation of the impedance-matching network is desired.

For series transmission lines, "tee" or "pi" lumped-element equivalent circuits can be genrated in order to predict performance or to make an approximate transformation [29]. The impedance matrix for a "tee" equivalent circuit is

$$Z = \begin{bmatrix} Z_a + Z_c & Z_c \\ Z_c & Z_b + Z_c \end{bmatrix} = \begin{bmatrix} Z_0 \cot\theta & Z_0 \csc\theta \\ Z_0 \csc\theta & Z_0 \cot\theta \end{bmatrix} \qquad (3\text{-}25)$$

and for a "pi" equivalent circuit it is

$$Y = \begin{bmatrix} Y_1 + Y_3 & -Y_3 \\ Y_2 & Y_2 + Y_3 \end{bmatrix} = \begin{bmatrix} Y_0 \cot\theta & -Y_0 \csc\theta \\ -Y_0 \csc\theta & Y_0 \cot\theta \end{bmatrix} \qquad (3\text{-}26)$$

Therefore, these networks are

$$Z_A = Z_B = j \, Z_0 \, \text{TAN} \, \frac{\theta}{2}$$

$$Z_C = -j \, Z_0/_{\text{SIN}\,\theta}$$

$$Y_1 = Y_2 = j \, Y_0 \, \text{TAN} \, \frac{\theta}{2}$$

$$Y_3 = -j \, Y_0/_{\text{SIN}\,\theta}$$

These approximations are accurate for θ values less than 45 degrees. For example, a series inductor of 1.5nH has an inductive reactance of $+j25.4$ ohms at 2700MHz. A section of transmission line with $Z_0 = 120$ ohms and $\theta = 12$ degrees has the same equivalent series reactance:

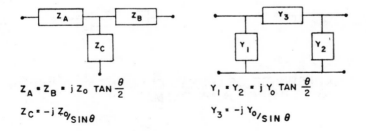

$$Z_L = 2\pi F L$$
$$= +j.25\,\Omega$$

The approximation can be seen graphically in Figure 3-35, which shows the impedance response of a series inductor and a length of transmission line modeled in a series inductor. However, for longer line lengths (>45 degrees) the approximations become increasingly inaccurate.

98

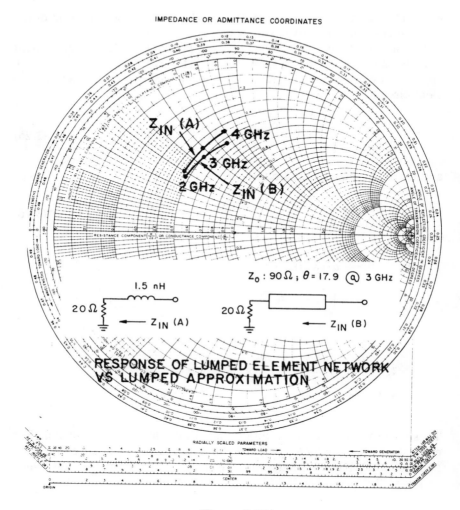

IMPEDANCE OR ADMITTANCE COORDINATES

$Z_0 : 90\,\Omega\; ;\; \theta = 17.9$ @ 3 GHz

RESPONSE OF LUMPED ELEMENT NETWORK
VS LUMPED APPROXIMATION

Figure 3-35

Another special case of the terminated transmission line occurs where
$\theta = 90$ degrees and the line is terminated in an impedance Z_L. Using the
general form:

$$Z_{in} = Z_0 \left[\frac{Z_L\cos\theta + jZ_0\sin\theta}{Z_0\cos\theta + jZ_L\sin\theta} \right] \qquad (3\text{-}27a)$$

and when $\theta = 90°$

$$Z_{in} = Z_0 \left[\frac{Z_L(0) + jZ_0(1)}{Z_0(0) + jZ_L(1)} \right] = \frac{Z_0^2}{Z_L} \qquad (3\text{-}27b)$$

which gives the results for the classic quarterwave matching section. A Wilkinson split-tee power divider is an example of how the quarterwave matching section can be used effectively [30]. Each 50-ohm output impedance is transformed through 70.7 ohms and $\theta = 90$ degrees to 100 ohms. The two 100-ohm arms are then combined in parallel to yield an input impedance of 50 ohms as shown in Figure 3-36.

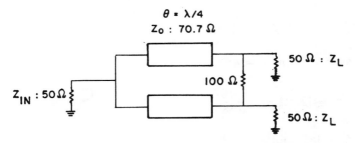

WILKINSON SPLIT-TEE COMBINER

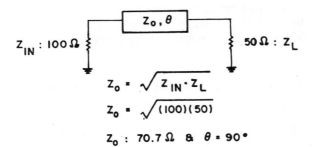

λ/4 WAVE MATCHING FOR THE EVEN
MODE ANALYSIS OF THE WILKINSON

Figure 3-36

Table 3-9 summarizes the approximations and special cases for trans-mission line matching. As a working example, the lumped-element input and output impedance-matching networks derived earlier will be con-verted to semi-lumped and distributed line equivalents. The lumped-element circuit is given as Figure 3-37. Working from left to right, the three shunt-mounted capacitors can be configured as open-circuited shunt stubs that are less than one quarter-wavelength. In practice, stub lengths greater than 65 degrees at the highest operating frequency should be avoided because the tangent function increases very quickly between 65 and 90 degrees. The effect is that longer stub lengths yield poorly re-peatable performance. The effective reactances are calculated:

$$Z_{stub} = \frac{-jZ_0}{\tan\theta} \qquad\qquad (3\text{-}28a)$$

$$(1) \quad Z_1 = -j50\Omega = \frac{-j50}{\tan 45°}$$

where $Z_0 = 15\Omega$ and $\theta = 45°$ $\qquad\qquad (3\text{-}28b)$

$$(2) \quad Z_2 = -j12.5\Omega = \frac{-j18}{\tan 55°}$$

where $Z_0 = 18\Omega$ and $\theta = 53°$

However, for easy physical implementation of these elements in the microstrip format, and for conserving circuit real estate, a practical com-promise is to build a physically symmetrical network with shunt and series transmission lines. Thus, the shunt reactances will be built as two parallel stubs:

$$Z_1 = -j50\Omega = Z_A \| Z_B, \text{ therefore } Z_A = Z_B = -j100\Omega \qquad (3\text{-}29a)$$

$$Z_2 = -12.5\Omega = Z_C \| Z_D, \text{ therefore } Z_C = Z_D = -j25\Omega \qquad (3\text{-}29b)$$

The microstrip stub that is closest to the device and offers the lowest reactance thus reaches the limit of practical usefulness. With a 60-degree limit imposed on the electrical length, Z_0 must be

$$-j2.5\Omega = \frac{-jZ_0}{\tan\theta} \qquad\qquad (3\text{-}30a)$$

and therefore

$$Z_C = 2.5 \tan\theta = 2.5 \tan 65°$$
$$Z_D = 5.36\Omega \qquad\qquad\qquad (3\text{-}30b)$$

Table 3-9

LUMPED	DISTRIBUTED EQUIVALENT
SERIES L 	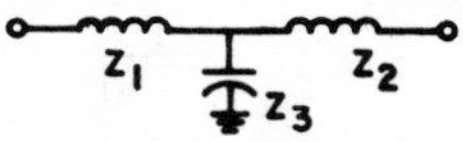 $Z_1 = Z_2 = jZ_0 TAN \frac{\theta}{2}$ $Z_3 = -jZ_0/_{SIN\ \theta}$
SERIES L 	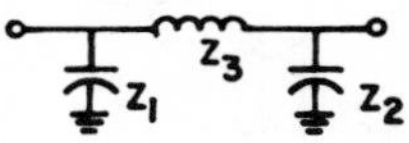 $Z_1 = Z_2 = -jZ_0 / TAN \frac{\theta}{2}$ $Z_3 = +jZ_0 SIN\ \theta$
SERIES 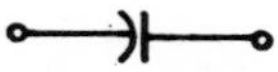	NOT PRACTICAL FOR VALUES OF C >.5 pF
SHUNT L 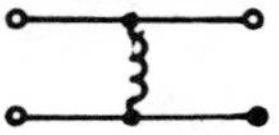	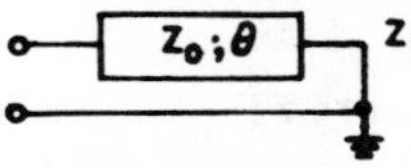 $Z_{IN} = +jZ_0 TAN\ \theta$
SHUNT C 	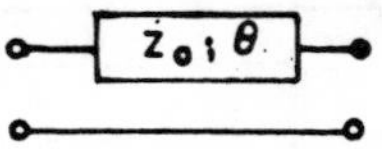 $Z_{IN} = \dfrac{-jZ_0}{TAN\ \theta}$
SERIES RESONANCE 	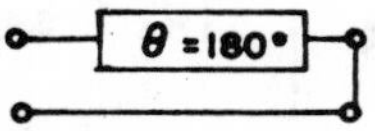 $\theta = 180°$ $Z_{IN} = j Z_0 TAN\ \theta$
PARALLEL RESONANCE 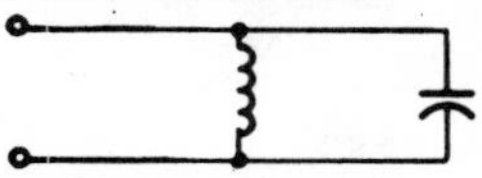	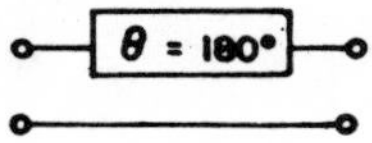 $\theta = 180°$ $Z_{IN} = \dfrac{-jZ_0}{TAN\ \theta}$

The useful range of characteristic impedances for the microstrip format runs from about 12 to 15 ohms for wide stubs to about 110 to 150 ohms for high-impedance narrow lines. Material selection generally determines these choices. Placing two stubs in parallel at the junction generally will yield a suitable implementation for the desired low reactance. However, in this case, the impedances are extremely low because the transistor manufacturer did not provide internal impedance-matching on the input of the device. Placing two stubs in parallel at the junction yields $Z_0 = 10.7$ and $\theta = 65$ degrees, and again the value of 10.7 ohms is too low for practical line widths. In this case it is best to distribute as much capacitance as possible and provide the remaining capacitance with a shunt-mounted chip capacitor. The equivalent is

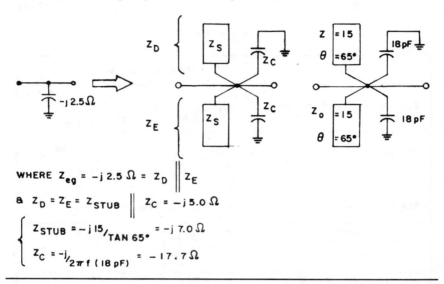

WHERE $Z_{eq} = -j\,2.5\ \Omega = Z_D \parallel Z_E$

& $Z_D = Z_E = Z_{STUB} \parallel Z_C = -j\,5.0\ \Omega$

$$\begin{cases} Z_{STUB} = -j\,15/_{TAN\ 65^\circ} = -j\,7.0\ \Omega \\ Z_C = -j/_{2\pi f\ (18\,pF)} = -17.7\ \Omega \end{cases}$$

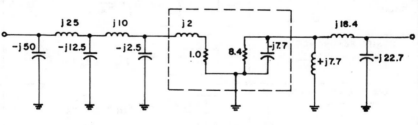

PROTOTYPE REACTANCE VALUES

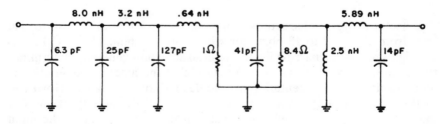

PROTOTYPE COMPONENT VALUES @ 500 MHz

Figure 3-37

The two inductive reactances in series between the three capacitive stubs can be transformed to the transmission line equivalents by the approximation:

WHERE

$$Z_1 = Z_2 = + j Z_0 \text{ TAN } \frac{\theta}{2}$$

$$Z_3 = \frac{-j Z_0}{\text{SIN } \theta}$$

The first element in the output circuit is a shorted length of transmission line (less than 90 degrees) that is used to imitate an inductance. This element is used to resonate the capacitance of the output. Note that collector matching external to the device is largely useless for frequencies above 500MHz because the series inductance contributed by the collector bond wires and the package lead inductances tend to bring the internal parallel RC network through resonance and into the realm of inductive reactance. Collector matching at the higher frequencies must be accomplished at the transistor die level in order to be effective. Beyond 3.0 GHz this, too, becomes a challenge. The first step in the design is to convert the output series equivalent impedance to the parallel equivalent, resonate the output capacitance with a shorted inductive stub, and then build a single-element structure in the same way as on the input:

$$Z_{D1} = Z_{D2} = j15.4 = j25 \text{ TAN } \theta \longrightarrow \theta = \text{TAN}^{-1} \left[\frac{15.4}{25} \right] = 31.6°$$

$$Z_E = 2(j Z_0 \text{ TAN } \theta/2) = j13.4 \longrightarrow Z_0 = 50, \theta = 15.3°$$

$$Z_F = -j18.7 = -j \frac{Z_0}{\text{TAN } \theta} \longrightarrow Z_0 = 20, \theta = 47°$$

The final circuit schematic for the transmission line and the microstrip format is shown in Figure 3-38 as a first-cut layout without the bias elements that are shown in Figure 3-39. Plots of input and output impedance for the lumped-element prototype and the transmission line equivalent are shown in Figures 3-40 and 3-41.

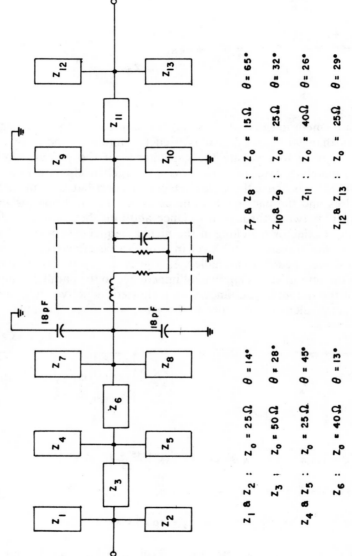

IMPEDANCE MATCHING WITH TRANSMISSION LINE ELEMENTS

Z_1 & Z_2 : $Z_0 = 25\,\Omega$ $\theta = 14°$

Z_3 : $Z_0 = 50\,\Omega$ $\theta = 28°$

Z_4 & Z_5 : $Z_0 = 25\,\Omega$ $\theta = 45°$

Z_6 : $Z_0 = 40\,\Omega$ $\theta = 13°$

Z_7 & Z_8 : $Z_0 = 15\,\Omega$ $\theta = 65°$

Z_{10} & Z_9 : $Z_0 = 25\,\Omega$ $\theta = 32°$

Z_{11} : $Z_0 = 40\,\Omega$ $\theta = 26°$

Z_{12} & Z_{13} : $Z_0 = 25\,\Omega$ $\theta = 29°$

Figure 3-38

BASELINE MATCHING
CIRCUIT LAYOUT

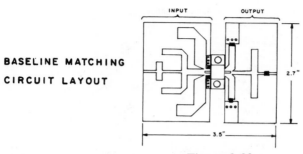

Figure 3-39

IMPEDANCE OR ADMITTANCE COORDINATES

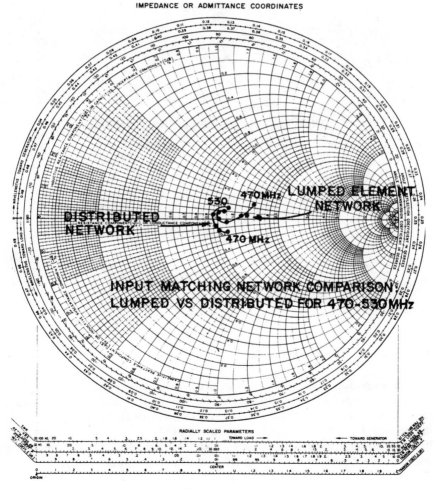

Figure 3-40

IMPEDANCE OR ADMITTANCE COORDINATES

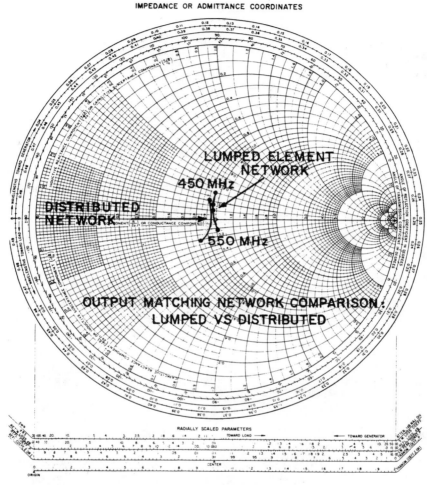

Figure 3-41

3.2.4 Class C Bias Components

The amplifier is biased by feeding the appropriate dc voltage to the input and output of the device as in Figure 3-42. A common procedure is to feed a positive collector potential (V_{cc}) through quarterwave high-impedance lines to the collector terminal of the device. An RF shorted, high-impedance quarterwave line in shunt with the collector matching will have negligible effect on the RF circuit performance. The input (base or emitter) is returned to dc ground in the same manner. In the common-base mode there is already enough parasitic resistance in the base-emitter path so that no series resistance need be added. In fact, for very high power devices, a very low value high-Q RF choke might provide higher stage gain because it has less of a tendency to reverse-bias the input than thin printed lines. No blocking capacitor is used on the input.

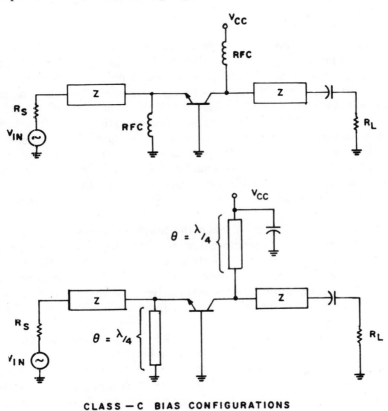

CLASS – C BIAS CONFIGURATIONS

Figure 3-42

3.3 DESIGN CONSIDERATIONS

Accurate impedance matching from the input and output of the power transistor represents only a small part of the overall amplifier design. Measurements of spurious output signals, junction temperatures, and junction temperature distribution are extremely important in the design phase. Complete knowledge of amplifier characteristics and behavior in the presence of load impedances that might be encountered in actual operation could prevent potential reliability problems. This section describes some of those characteristics, all of which should be considered during device and circuit characterization.

3.3.1 Amplifier Stability

Spurious oscillations in an amplifier are common in class C design. In general, a spurious oscillation is the result of an interaction between input and output elements, and can be caused by a myriad of interdependent variables for different operating conditions. Needless to say, causes and cures are difficult to identify [31,32]. The common-base configuration is particularly prone because of regenerative effects due to the common lead inductances in the base circuit. Because of its emitter-into-base-into-collector vertical diffusion profiles, the base and emitter contacts are accessed on the upper surface of the die. Wire bonds must provide the grounding path from the base contact on the transistor chip to the package ground. These are shared inductances on the input and output.

Instabilities generally exist when the amplifier demonstrates one or more of the following characteristics: power output "snap-on" as RF drive is slowly applied; an abrupt drop in power output as RF drive is increased; abrupt changes in power output or efficiency with frequency; or hard saturating output power levels. Because there are numerous modes of input-output interaction, many involving the nonlinear device characteristics, there are numerous causes of spurious oscillations. A mature transistor device design and good RF circuit design and construction techniques will usually lead to a stable amplifier.

One of the most common forms of instability is a low frequency instability, which can be seen on a spectrum analyzer as a set of spectral lines flanking the carrier frequency. Because the low frequency gain of an RF device is extremely high, tuned circuits or feedback paths for these frequencies must be avoided, and bypassing at low frequencies is critical. RF chokes with various values of bypass capacitors ranging from 100pF to 10μF may be necessary in the dc collector feed. A lossy ferrite bead can also be slipped onto an RF choke at the ground end. Another solution involves reducing the overall Q of the input network or making the output Q higher.

Parametric oscillations arise from the nonlinear varactor effect of the collector-base output capacitance and are not a feedback phenomenon.

This instability is manifested as a subharmonic of the carrier frequency, for example, as an $F_0/2$ oscillation. It is generally more device-oriented and can be avoided by using a well engineered, mature device design. Parametric oscillations may arise when operating a device well below its rated collector voltage, where the C_{ob} tends to increase and the RF voltage swing tends to drive the device into a parametric mode. The schematic of Figure 3-43 indicates bypassing elements necessary to maintain stability for a particular design.

3.3.2 Balanced Amplifier Designs

Balanced amplifier designs that use a dual-lead transistor in a push-pull configuration can offer significant advantages over equivalent single-ended designs [33,34,35,36]. A dual-lead balanced transistor contains two transistor chip groups operated 180 degrees out of phase about an internal virtual ground. The virtual ground is the selling point of this configuration because it reduces the common mode inductance that is a major source of instability. Also the mounting of the transistor becomes less critical and presents an opportunity for some unique circuit configurations. Additionally, input and output impedances are effectively increased. A dual-lead package that is operated as two isolated single-ended transistors presents terminal impedances that are half that of the lead-to-lead balanced impedances. Balanced impedances are four times the value of the single-ended impedance for a device with equal emitter periphery. Figure 3-44 outlines this comparison. The balanced configuration provides for unique impedance-matching configurations within the device and in the external matching networks. Shunt capacitances can be mounted line-to-line. These can be particularly attractive as a space saving feature in an impedance-matching network at UHF frequencies using chip capacitors. The advantages of balanced operation can only be exploited if the 180-degree line-to-line phase relationship is maintained. This necessarily leads to the design of an external balun (balanced-to-unbalanced transformer), which must be included as a component of the external impedance-matching networks and tends to make the external circuitry more complicated [37].

3.3.3 Design for Minimum Junction Temperatures

RF power transistors in increasingly complex systems must provide adequate performance under harsh environmental conditions. The trend

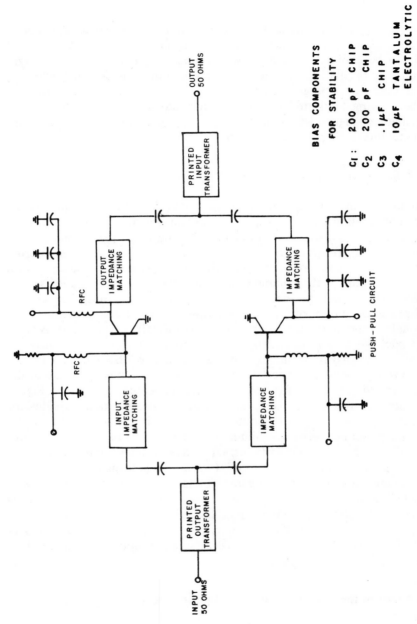

Figure 3-43

toward higher frequencies, higher power density devices, higher packaging densities, and increased bandwidths can often inadvertantly lead to higher dissipation in the device. Because many of the failure mechanisms of the device are related to temperature, it is important to maintain low and evenly distributed junction temperatures within the device.

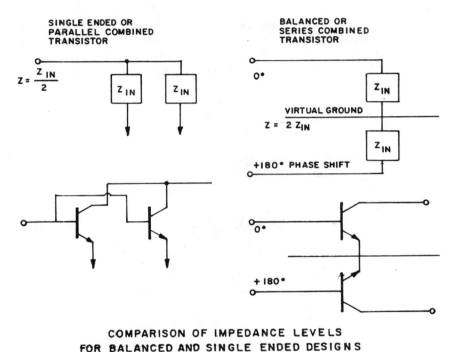

**COMPARISON OF IMPEDANCE LEVELS
FOR BALANCED AND SINGLE ENDED DESIGNS**

Figure 3-44

There are two fundamentally sound methods for measuring junction temperatures so that the effective device thermal resistance can be calculated. The most commonly used method is the infrared scanning technique, which uses an infrared detector in conjunction with an optical microscope. The microscope can be focused on the surface of the transistor die and the emanating infrared energy can be measured as a detected voltage. The detected voltage can be compared with a calibrated reference curve and a peak transient temperature can thus be measured. Infrared scanners can focus on image areas as small as 1 mil in diameter and detect thermal transients as fast as 5μs. Using this technique, a designer can determine how the hot-spot junction temperatures vary under different operating conditions and how these conditions can be avoided. A mature device design can be set apart from a relatively fragile design by measurement of such parameters as junction temperature (T_j) distribution across the surface of the transistor die [38]. A well designed transistor-amplifier stage will minimize the T_j spread. The scanning method necessarily requires the use of uncovered devices and time-consuming calibration and measurement. Figure 3-45 shows the functional parts of an infrared scanner [39] and Figure 3-46 is a photograph of the oscilloscope video display for a device operating in a pulsed mode.

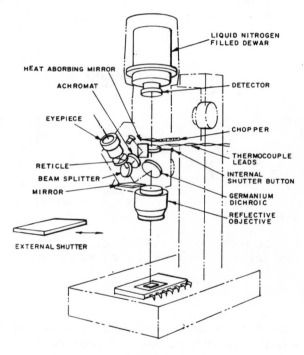

Figure 3-45

The second technique involves the measurement of the temperature variant base-emitter diode voltage (V_{be}) of the transistor. Only the average temperature can be calculated by measuring V_{be}. This method cannot measure the development of hot-spots, which are pervasive in large multicell devices. It is rather a more subjective test for large high power devices. In cases where the infra-red scanning technique cannot be used (e.g., for covered devices, if the cell area is too small, or if metallized regions over the active area), the V_{be} measurement method can supply a baseline comparison. In this method, T_j is calculated from the bipolar transistor relation for the emitter current (I_e), where I_e is dependent on geometry, material constants, and external parameters:

$$T_j = \cfrac{qV_{BE}}{k\ln\left[\dfrac{J_E N_B}{A_e q n_i^2 D_B} + 1\right]}$$

(3-31)

where

V_{BE}	= base-emitter voltage	D_B, N_B, n_i	= semiconductor material
q	= charge constant		constants
I_E	= emitter current	A_e	= emitter area
		k	= Boltzmann's constant

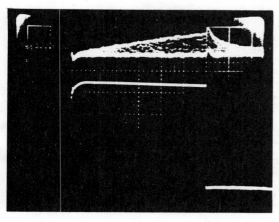

IR MICROSCOPE
VIDEO OUTPUT
HOTTEST CELL: 104°C

DETECTED
OUTPUT PULSE
20 µs/cm HORIZONTAL

FREQ	P_{IN}	P_{OUT}	I_C	V_{CC}	y	G_P	R.L.	T_{SINK}	θ_{j-s}
850	20W	148W	7.60	34V	57%	8.7 dB	19 dB	35°C	0.55°C/W

CELL #	1	2	3	4	5	6	7	8	9	10	11	12	13	14	15	16
T_j	104°	101°	99°	101°	92°	92°	92°	101°	95°	95°	95°	95°	99°	95°	95°	101°

CELL #	1	2	3	4	5	6	7	8	9	10	11	12	13	14	15	16
T_j	101°	99°	99°	99°	99°	101°	101°	101°	99°	101°	104°	104°	101°	101°	99°	101°

Figure 3-46

A meaningful parameter used in predicting T_j under similar yet slightly different conditions is the thermal resistance (θ_{jc}) of the device. The value θ_{jc} is actually a sum of three different terms, all accounting for the common construction techniques of bipolar devices as shown in Figure 3-47. Individual θ_{jc} values are dependent on geometry, materials, and three-dimensional heat flow; calculations for such are documented [40,41]. Junction temperature is given by

$$T_j = T_A + P_D(\theta_{jc}) \tag{3-32}$$

where

T_a = ambient sink temperature
P_d = peak dissipated power

$\theta_{jc} = \theta_{chip} + \theta_{eutectic} + \theta_{package}$

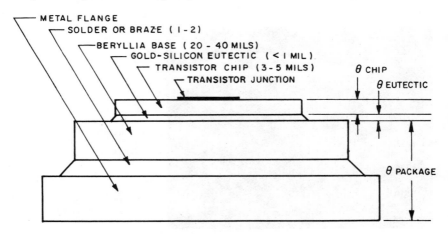

$\theta_{jc} = \theta$ CHIP + θ EUTECTIC + θ PACKAGE

COMPONENTS OF POWER TRANSISTOR THERMAL RESISTANCE

Figure 3-47

The actual semiconductor junctions are located on the surface of the transistor chip, about 1 micron deep, and the immediate environment below the chip has a tremendous thermal capacity. In terms of actual device operation, this means that the transient θ_{jc} exhibits a thermal time constant. The thermal time constant can be exploited when the transistor need only be operated for short pulse operation. Because the peak junction temperature does not respond instantaneously due to of the large thermal capacity, higher dissipated power than nominal can be experienced without stress on the device. Figure 3-48 illustrates measured temperature as a function of time for a 250µs RF pulse. In general, for the same semiconductor material type, as the die thickness increases, the thermal time constant increases, and the effective θ_{jc} is decreased. For example, higher frequency 3-GHz devices heat up quickly, having a lower thermal time constant than the larger and physically thicker UHF devices. Measured data show that operating RF pulse widths smaller than the thermal time constant of the device leads to lower values for T_j, while for long pulse operation, on the order of 2 to 3 thermal time constants, the temperature of the device approaches the CW condition. Exploiting the characteristics of the thermal time constant, especially in the case of very short RF pulse operation, the peak power output of the device can be increased as much as twice that of the CW condition while maintaining the same equivalent peak T_j [42].

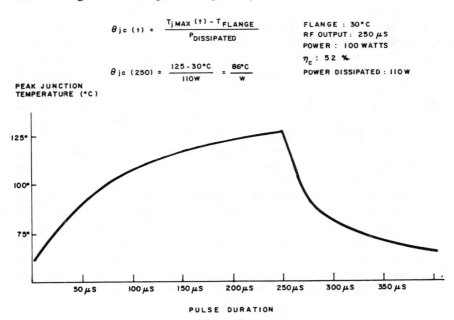

$$\theta_{jc}(t) = \frac{T_{j\,MAX}(t) - T_{FLANGE}}{P_{DISSIPATED}}$$

$$\theta_{jc}(250) = \frac{125 - 30\,°C}{110W} = \frac{86\,°C}{W}$$

FLANGE : 30°C
RF OUTPUT : 250 µS
POWER : 100 WATTS
η_c : 52 %
POWER DISSIPATED : 110 W

PEAK JUNCTION TEMPERATURE (°C)

PULSE DURATION

Figure 3-48

3.3.4 Collector Load Contouring

Device manufacturers normally rate their devices at the device's maximum power output capability. The device rating needs to be adjusted to ensure low temperature operation, and this is accomplished at the expense of power output. The design objective is to maintain reliable performance while supplying minimal power output along with allowing for nominal system and production tolerances. The effects can be realistically approached in the amplifier design phase with the aid of a generic set of curves for a particular transistor, caled the *collector load contours*.

Collector load contours are diagrams drawn on a polar impedance plot that show device characteristics varying as a function of the collector load impedances [43,44,45]. For a given frequency, collector supply voltage, and RF drive level, varying the complex impedance at the output of the device will vary the power output, efficiency, and junction temperatures within. These parameters can be plotted on a Smith chart with an accurately calibrated system. Figure 3-49 shows a Smith chart representation of a collector load contour. Given a nominal level for V_{cc}, P_{in}, and frequency, we find that there is a single impedance for which the output power is maximized. This represents the maximum power output that can be achieved unless the collector voltage drive or the frequency, are altered. This point represents a single impedance, and any deviation from this impedance point results in a lower level of output power than was achieved at the P_{max} point. With the power output reduced by an arbitrary level, such as 0.5dB below the P_{max} point, there are an infinite number of impedances that will result in this power level. These lower power points represent a concentric locus about the P_{max} point. Any one of these loci represents a contour on which any impedance will derive the same power output from the device.

Another parameter which can be related to the contour is the peak T_j. For devices without internal collector matching, or for devices with internal collector matching that are resonated above the operating band, the junction temperatures are higher on the lower impedance levels and lower on the higher impedance levels. By observing the effects that some impedances have on junction temperatures, contours of T_j can be plotted in the same manner as power output is plotted on the Smith chart. Two important facts emerge from the concept of load contouring. First, a device must be capable of more power output than required from the single-stage amplifier. Second, a matching network can be selected that will produce a lower junction temperature for an equivalent mismatch as another impedance. This is represented in Figure 3-49(b) by shifting the

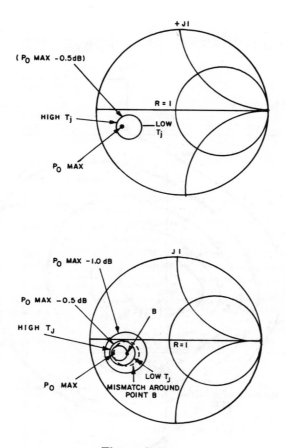

Figure 3-49

matching point from P_{max} to B. Actual load contours for a standard packaged device operated at 900MHz in a pulsed made are shown in Figures 3-50 through 3-55. Note that power output, efficiency, peak junction temperature, and insertion phase are all plotted for this operating condition. A shift in frequency, drive power, or collector voltage will shift these curves to a different location. The use of collector load contouring techniques to examine the interaction of load impedances with the operating parameters of the transistor permits optimization of these parameters and provides a practical means of establishing measurable performance specifications for the device.

CONTOURS OF PEAK POWER OUTPUT

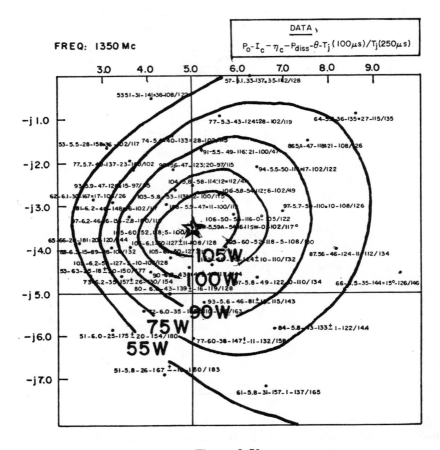

Figure 3-50

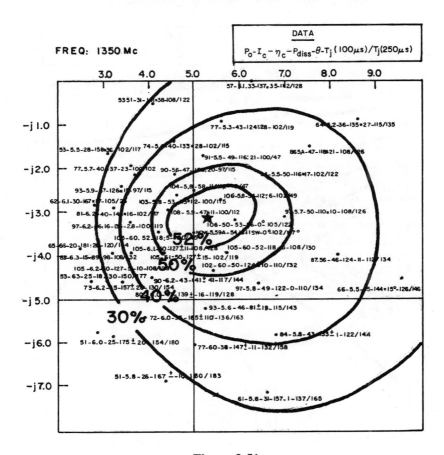

Figure 3-51

120

CONTOURS OF COLLECTOR CURRENT

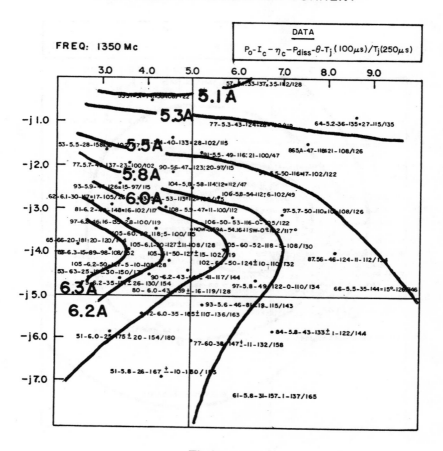

Figure 3-52

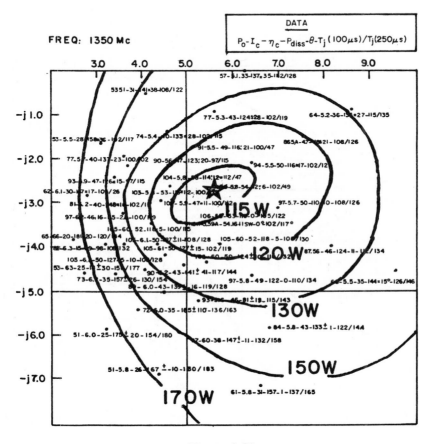

Figure 3-53

CONTOURS OF RELATIVE INSERTION PHASE

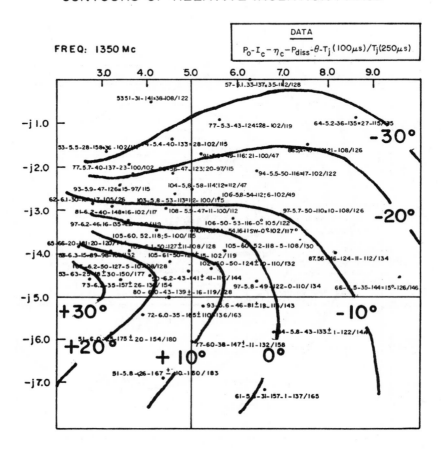

Figure 3-54

CONTOURS OF PEAK JUNCTION TEMPERATURE

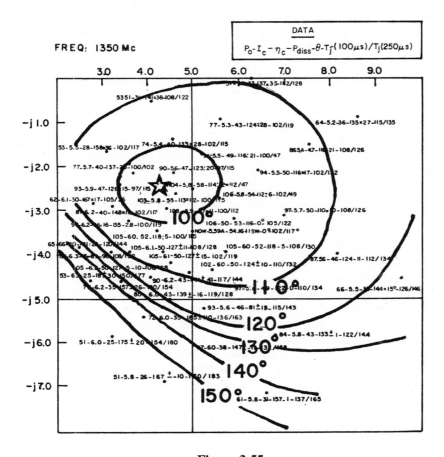

Figure 3-55

3.3.5 Junction Temperature Effects and Load Mismatch

It is a time-consuming procedure to perform load-pull characterization without introducing new variables. However, a major parameter under scrutiny is the hot-spot junction temperature and the way in which the temperature is affected by variations in load mismatch. A complication arises because the hot-spot temperatue may vary from cell to cell within a multicell device for different phases of output mismatch of the same magnitude. Multicell devices are complex hybrid circuits with multiple-wire bonds, chip components, and microstrip transmission lines. Load impedances that are measured at the terminals of the device are transformed to quite lower impedances at the transistor die level. Coupled with the fact that multicell devices tend to be physically large, significant changes in RF drive distribution, and consequently temperature, can occur for different phases of load mismatch of the same magnitude. Therefore, P_{out}, T_{jmax}, and efficiency are not only functions of the complex load impedance, but so too can the junction temperature cell-to-cell differential vary significantly with impedance changes. At the low impedances near the die, two specifically selected 180-degree relative phase mismatches of equal magnitude will present load matches that are higher and lower than the value for a conjugate match. In other words, there is a state of high voltage/low current and one of low voltage/high current at the transistor die level. Figure 3-56 represents these conditions for a multicell device operated under a 2:1 mismatch around its point of maximum power output. In this plot, each of the device's 32 cell-junction temperatures have been plotted for 10 different phases of 2:1 output VSWR. The device is operated at 800MHz and below the resonant frequency of the collector match. Note that the temperature distribution *versus* mismatch angle fluctuates widely, and hot-spot areas for some impedances appear benign at others. Plots 4, 5, 6, and 7 in Figure 3-56 point to regions in the impedance plane where the cells within the center of the chip show a tendency to develop hot spots, while plots 1, 2, 9, and 10 point to load impedances that cause cells in close proximity to the package side walls to develop hot spots. Figure 3-57 helps to identify cell position with the measured junction temperature.

125

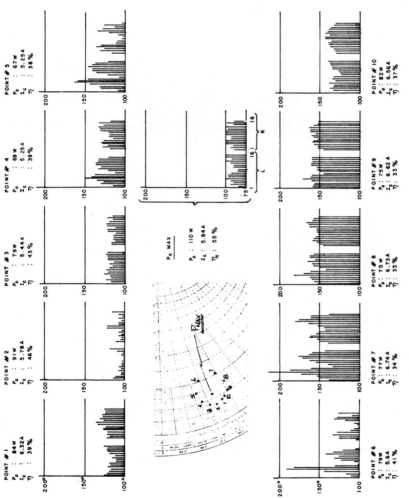

JUNCTION TEMPERATURE DIFFERENTIALS FOR SCATTERE PHASES OF 2:1 VSWR

Figure 3-56

126

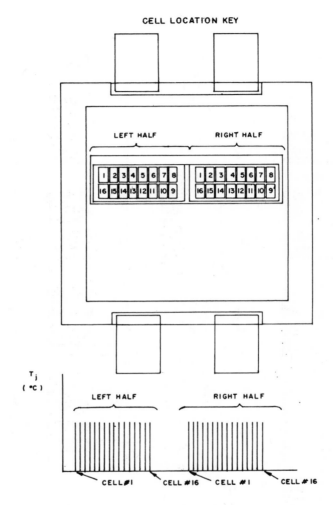

CELL LOCATION KEY

Figure 3-57

REFERENCES

1. C. Snapp, "Bipolars Quietly Dominate," *Microwave Systems News,* Nov. 1979, pp. 45–67.
2. W. Poole, "S-Band Transistors for Radar Applications," *Microwave Journal,* March 1983, pp. 85–90.
3. G. Schreyer, "High Performance S-Band Radar Transistor," *Microwave Journal,* March 1981, pp. 63–66.
4. R. Basset and M. McCombs, "Production Bipolars Edge Out FET's at 4GHz, *Microwaves,* Feb. 1981, pp. 43–49.
5. J. Johnson and D. Wisherd, "Solid-State Power for L-Band Radar," *Microwave Journal,* Aug. 1980.
6. R. Allison, "Silicon Bipolar Microwave Power Transistors," *IEEE Trans. Microwave Theory Tech.,* vol. MTT-27, no. 5, May 1979, pp. 415–422.
7. Microwave Semiconductor Corporation "Pulsed Power Transistor: 1 HP at 1GHz," *Microwave Journal,* June 1979, p.34.
8. C. Snapp, "Silicon Bipolar Transistors and Integrated Circuits Continue to Grow," *Microwave Systems Designers Handbook,* pp. 32–41.
9. IEEE MTT, Boston Chapter, Seminar on Microwave Devices and Applications, April 1975.
10. T. T. Ha, *Solid-State Microwave Amplifier Design,* New York, John Wiley and Sons, 1981, pp. 105–110.
11. R. Bailey, "Large Signal Non-Linear Analysis of High Power Transistors," *IEEE Trans. Electron Dev.,* vol. ED-17, no. 2, Feb. 1970, 108–119.
12. Ha, *op. cit.,* pp. 105–110.
13. O. Pitzalis and R. Gilson, "Broadband Microwave Class-C Transistor Amplifiers, *IEEE Trans. Microwave Theory Tech.,* vol. MTT-21, no. 11, Nov. 1973, pp. 660–668.
14. H. Cooke, "Microwave Transistors: Theory and Design," *Proc. IEEE,* vol. 59, no. 8, Aug. 1971, pp. 1163–1181.
15. J. Curtis, "Modeling Transistors for Improved Performance," *Microwaves,* Feb. 1975, pp.33–43.
16. C. Nelson and G. Stuart, "High Efficiency Amplifier Study," Final Report: RADC-TR-78-266, Jan. 1979.
17. Pitzalis and Gilson, *loc. cit.,* pp. 660–668.
18. Cooke, *loc. cit.,* pp. 1163–1181.
19. I. Getreau, "Modeling the Bipolar Transistor," Tektronix, Inc., Beaverton, OR, March 1976.

20. R. Pucel, "An Analysis of the Lead Inductances of a devel Transistor," Raytheon Research Division, Technical Memorandum No. T912, Feb. 1972.
21. R. Hale, Engineering Manager, M/A-COM PHI, Torrance, CA, private communication, Jan. 1985.
22. E. Joly, President, Rhode Island Electronic Ceramics, Greenville, RI, private communication, Jan. 1985.
23. *Ibid.*
24. H. A. Wheeler, "Transmission Line Properties of a Strip on a Dielectric Sheet on a Plane," *IEEE Trans. Microwave Theory Tech.*, vol. MTT-25, no. 8, Aug. 1977, pp. 631–647.
25. Ha, *op. cit.*, pp. 289–295.
26. Rogers Corporation, "Width and Effective Dielectric Constant Data for Design of Microstrip Transmission Lines on Various Thicknesses, Types, and Claddings of RT/Duroid Microwave Laminates," RT3.1.2, Chandler, AZ, April 1982.
27. T. C. Edwards, *Foundations for Microstrip Circuit Design,* New York, John Wiley and Sons, 1981, p. 45.
28. H. Howe, *Stripline Circuit Design,* Dedham, MA, Artech House, 1974, pp. 1–32.
29. Brown, Sharpe, Hughes, and Post, *Lines, Waves, and Antennas: The Transmission of Electric Energy,* The Ronald Press Company, pp. 147–176.
30. E. Wilkinson, "An N-Way Hybrid Power Divider" *IEEE Trans. Microwave Theory Tech.,* vol. MTT-8, no. 1, Jan. 1960, pp. 116–118.
31. H. O. Granberg, "Good RF Construction Practices and Techniques," *RF Design,* Sept./Oct. 1980, pp. 51–59.
32. N. Sokal, "Parasitic Oscillations in Solid-State RF Power Amplifiers," *RF Design,* Nov./Dec. 1980, pp. 32–36.
33. L. Max, "Balanced Transistors: A New Option for RF Design," *Microwaves,* June 1977, pp. 42–46.
34. L. Max, "Apply Wideband Techniques to Balanced Amplifiers," *Microwaves,* April 1980, pp. 83–88.
35. D. Wisherd, "Apply Microwave Methods to Balanced Amplifiers," Microwaves, July 1980, pp. 54–62.
36. J. Johnson, "A Look Inside Those Integrated Two-Chip Amps," *Microwaves,* Feb. 1980, pp. 54–59.
37. R. Basset, "Three Balun Designs for Push-Pull Amplifiers," *Microwaves,* July 1980, pp. 47–52.
38. M. Flahie, "Reliability—The Long and the Short of It," *Microwaves,* 1972.

39. J. Anslow, G. Baxter, and S. Brouillette, "Thermal Resistance of Microelectronic Packages," Report No. RADC-TR-77-321, Oct. 1977.
40. R. Linsted and R. Surty, "Steady-State Temperatures of Semiconductor Chips," *IEEE Trans. Electron Devices,* vol. ED-19, no. 1, Jan. 1972, pp. 41–44.
41. H. Cooke, "FET's and Bipolars Differ When the Going Gets Hot," *Microwaves,* Feb. 1978, pp. 55–61.
42. E. Brookner, "Present and Future Trends in Radar Systems and Components," *Radar Technology,* Dedham, MA, Artech House, 1973, pp. 327.
43. R. Naster and W. Perkins, "Solid-State Power Amplifiers for L-Band Phased Arrays," *Microwave Journal,* July 1975, pp. 56–59.
44. Pitzalis and Gilson, *loc. cit.,* pp. 660–668.
45. A. Rosen, D. Stevenson, and A. Presser, "Hybrid Integrated 10-Watt CW Broadband Power Source at S-Band," *IEEE J. Solid State Circuits,* vol. SC-4, no. 12, Dec. 1969.

Chapter 4

The Corporate Structure Amplifier

The *corporate structure* has been employed in the past in antenna feed networks to divide the power output of a single vacuum tube transmitter in a controlled manner in order to feed a large number of radiating elements in the transmission mode and to combine the signals from these elements during the reception interval.

The network provides a constant time delay between each radiating element and the common feed point. The network derives its name from the organizational chart of a large management chain in which the number of people reporting directly to any manager is small, typically five or six. The levels of management between any low-level work station and the highest level of management is a relatively small number. For example, with six people reporting to the next higher management level, an entire population of 55,986 employees could boast that there were less than five levels of management between them and the chief executive officer (the populations of the levels of management above the lowest group of 46,656 in this example are 7776, 1296, 216, 36, and 6 under control of one person).

This hierarchical structure is a very efficient organization of a large number of controlled elements with a relatively small number of superior elements in control of the lower elements. It has naturally been applied to the management of a wide variety of organization types including political, commercial, military, and religious. The analogy to the organization of a high-power transmitter comprised of a large number of low-power elements is the subject of this chapter. Some elements of the transmitter are included only because they are needed to control the action of lower level elements, which are directly coupled to the output load of the transmitter. These vital, but indirectly involved, components are analogous to the management elements of a commercial institution. An organization with a small number of management level elements is usually, but not always, more efficient than one with a higher population of managers.

There is a special case in which a CSA is made up of identical building blocks of unit amplifiers. It can be shown that this approach requires the minimum number of active elements to produce a given required output power. It has the property that the number of transistors driven by any transistor (fan-out ratio) is a constant, which is equal to the power gain of the unit amplifier. This *uniform corporate structure amplifier* (UCSA) has not been employed in any of the solid-state transmitters currently in service because of yield problems with the very high power rated output transistors used in early transmitter module designs.

The more usual type of CSA employs different unit amplifier designs for each level of the hierarchy, usually with a lower power rating and higher gain transistor at the input end, and higher power, lower gain devices at the output end.

The use of identical amplifier elements in the UCSA is based on a logic which says that a design should be used only after it has been carefully analyzed and tested. If a different type of design has any merit, then it should become the "standard." If the "standard" is a reliable element, then it might as well be used at all levels. This procedure also has the virtue that a high percentage of the engineering cost can be budgeted for the development of a single building block—including the mechanical, thermal, reliability, and manufacturing engineering, which are so important in a structure that is by its nature composed of a large number of active elements.

The next level of structural complexity in the CSA is a module, which is composed of a number of power gain elements. A module is a similar but more complex building block, which will provide more power output and usually more power gain, and will occupy more space and cost more than the single transistor gain cell. The CSA theorems apply equally well to each.

4.1 PROPERTIES OF THE CORPORATE STRUCTURE AMPLIFIER

A UCSA can produce a given required peak output power with fewer active elements than a transmitter in which a mix of different transistor types is used. The UCSA employs transistor types with the highest reliable power output rating from the set of devices available. Proofs for the following properties of the UCSA are given in Appendix A.

(a) A minimum number of unit amplifiers is needed to provide a given required output peak power when the "fan-out ratio" is equal to the unit amplifier power gain. Fan-out ratio is defined as the number of unit amplifiers driven by each unit amplifier.

(b) For transistor types with the same gain and different power output ratings, a CSA utilizing the highest output power devices from the above set will require a minimum number of transistors to produce a given required power output.

(c) When a given power output is developed by summing the output of two or more CSAs, the number of transistors used is the same as would be required in a single CSA with the same power output and power gain per stage.

(d) When a CSA is built using modules which contain more than one unit amplifier, the fan-out ratio between modules is equal to the module gain and will produce a given required output power with a minimum number of modules. If the module is also a CSA, the number of transistors needed to produce a given required power output for the modular CSA is identical to that required in a non-modular CSA composed of unit amplifiers which produce the same power output.

A block diagram of a UCSA utilizing 6–dB gain unit amplifiers is shown in Figure 4-1. Each unit amplifier drives four identical stages in this example. The output of the final 16 stages is combined to provide a single output. For phased array applications, the output stages would be preceded by phase shifters, and radiating elements of the array would replace the power combiner network. The overall power gain of the UCSA shown in Figure 4-1 is 18 dB. The fan-out ratio of 4 to 1 is equal to the power gain of each unit amplifier minus the losses associated with the power divider network. The amplifiers are usually operated in the class C mode and close to saturation. The amplifiers are relatively insensitive to changes in drive level because of the saturation effects.

4.2 ANALYSIS OF THE UNIFORM CSA

The following analysis applies to the UCSA. These models may also be useful for estimation and size comparison with the non-uniform CSA.

If the rank of a unit amplifier in a UCSA is defined as the number of unit amplifiers between its input and the input node of the UCSA, then the number of unit amplifiers in rank m with a fan-out ratio G is

$$N_m = G^m \tag{4-1}$$

In the example of Figure 4-1, $m = 2$ and $G = 4$ at the output rank. The overall gain of the UCSA is $G^{(m+1)}$; or, in dB notation, $G_t = (m + 1)G_{dB}$.

The total number of unit amplifiers in a UCSA with an output group of rank K is

$$N_t = \sum_{m=0}^{K} G^m = \frac{(G^{(K+1)} - 1)}{G - 1} \tag{4-2}$$

134

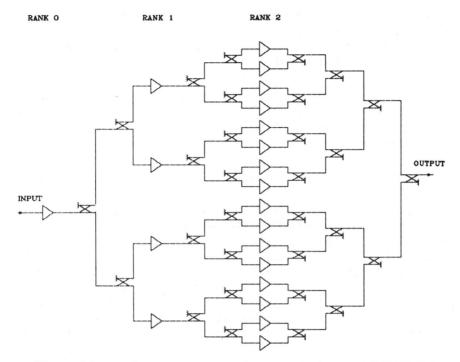

Figure 4-1   CORPORATE STRUCTURE AMPLIFIER WITH 6 dB UNIT AMPLIFIERS

If the power output of each unit amplifier is P_u watts, the power input to the final power combiner is $G^m P_u$. The UCSA output power is reduced by the amount of final power combiner loss. The useful power output is developed by the unit amplifiers in the output rank of the UCSA. All other unit amplifiers are required to provide drive power to the unit amplifiers in the output rank. These driver stages form a CSA of rank $m - 1$. If the overhead ratio is defined as the fraction of total CSA devices needed to drive the output rank to the number of unit amplifiers in the output rank, then the overhead ratio is

$$O = (1 - G^{-K})/(G - 1) \tag{4-3}$$

For the example of Figure 4-1, the overhead ratio is 5/16, or 0.3125. If the same number of output stages were used in a CSA with a fan-out ratio of 2:1, the overhead ratio would be 0.9375 and the 16 output amplifiers would require 15 driver amplifiers $(8 + 4 + 2 + 1)$. This shows the penalty of a low power gain unit amplifier on the overall cost and reliability of a CSA. The same output is obtained with 21 unit amplifiers with a 6 dB per stage power gain as 31 amplifiers with a 3 dB power gain. Ultimately, the same output could be provided by 17 amplifiers with a 16:1 fan-out ratio. Any increase in gain beyond 12 dB would result in excess drive power to the output rank. This is an impractical example, included to illustrate the effect of stage gain on the overhead ratio. At this time, typical stage power gain at L band and below is in the 6 to 8 dB range, and in the 3 to 5 dB range at S band. The gain usually decreases as power output per stage increases.

4.3 TRANSMITTER EFFICIENCY

The efficiency of a transmitter is defined as the ratio of the RF output to the dc and RF input power. High efficiency is desired in order to reduce the operating temperature of the semiconductors and associated cooling system costs. Efficiency is calculated from Eqs. (4-1) and (4-2) as

$$\eta_t = \frac{\eta_u L_c (G^{(K+1)} - G^K)}{(G^{(K+1)} - 1)} \tag{4-4}$$

where

η_u = Unit amplifier efficiency
L_c = Output circuit losses

In most cases, $G^{(K+1)} \gg 1$ and the following simple equation is valid

$$\eta_t = \eta_u L_c (1 - G^{-1}) \tag{4-5}$$

For example, for the UCSA illustrated in Figure 4-1, the efficiency of each unit amplifier is 60 percent. The gain of each stage is 6 dB and $K = 2$. The above equation gives an overall efficiency $\eta_t = 45.7$ percent. If the output power is held constant and the gain per stage decreased to 3 dB ($G = 2$), then K becomes 4 and the overall efficiency $\eta_t = 30.97$ percent. The variation of efficiency with unit amplifier gain and K is illustrated in Figure 4-2, where the overall transmitter efficiency is plotted against K for unit amplifier gains of 3, 6, and 9 dB with unit amplifier efficiency equal to 100 percent. In a real case, the unit amplifier efficiency is in the 40 to 65 percent range. The curves in Figure 4-2 can be used to estimate overall transmitter efficiency when the unit amplifier efficiency is known through multiplying the maximum efficiency by the unit amplifier efficiency and the output combiner loss factor L_c.

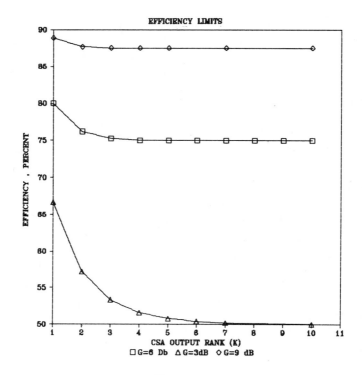

EFFICIENCY LIMITS

Figure 4-2

4.4 COMPARISON OF THE *PAVE PAWS* TRANSMITTER WITH A UCSA

The PAVE PAWS radar (AN/FPS-115) was built in 1978 at Otis Air Force Base on Cape Cod for the detection and tracking of submarine launched ballistic missiles. The selection of a solid-state transmitter for the radar was based on a trade-off analysis of the operation and maintenance cost as well as acquisition cost. The decision to employ bipolar transistors instead of vacuum tubes was a courageous one, which was proved valid by the rapid completion of the design, construction, and subsequent evaluation. The module used in the PAVE PAWS is shown in block diagram form in Figure 4-3. This module is clearly *not* a UCSA design. There was a good reason for departing from the UCSA approach, which was based on a study of the overall cost of acquisition. The specifications for the output stage devices were a challenge to the transistor manufacturer and it became apparent that there would be a large

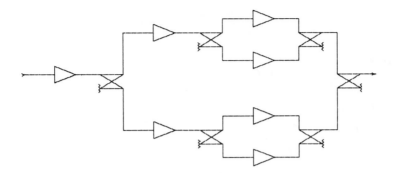

Figure 4-3   PAVE PAWS Transmitter Module Block Diagram

number of perfectly good devices that could not quite meet the demanding specifications for the output devices and yet be quite capable performers at a reduced power level. The overall cost of these modules was minimized by using the devices that would not be acceptable as output rank members as driver stages for that rank. This is a good example of a system in which all aspects of performance and cost were considered and the result was a viable compromise. It is interesting to compare the total number of devices needed to produce the PAVE PAWS output peak power in a UCSA with the design that was implemented. The PAVE PAWS employs a single (with redundant backup) driver module followed by a 1:56 power divider, which feeds 56 subarray power dividers. Each subarray power divider is a 1:32 which drives a group of 32 transceiver modules. For each array face, there are 1792 (32 × 56) transceiver modules. This is shown in the simplified system block diagram at the top of Figure 4-4.

If a UCSA design were implemented (shown at the bottom of Figure 4-4), the module would consist of a single device that drives four output stages. The output power of each module is the same as that used in the original design. The module gain is lower because only two gain stages are used. The overall power gain of the UCSA module, including combiner and other microwave losses, is 12 dB. The corresponding fan-out ratio is 1:16. This results in a subarray with only 16 modules *versus* the 32 used in the PAVE PAWS. A total of 112 subarray drivers are required and an equal number of phase shifters. The power rating of the phase shifters is somewhat higher than that used in the original design because of the lower power gain of the UCSA module. It requires a peak power handling capability of the order of 17.5W, which is not a very demanding requirement. The 112 subarray phase shifters are driven by a pair of 1:56

Comparison Of PAVE PAWS With
Equivalent UCSA Architecture

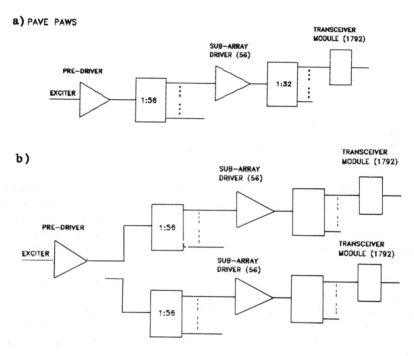

a) PAVE PAWS

b)

Figure 4-4

power dividers, which are identical to those used in the original design, although they are rated at a somewhat higher power level, again because of the low module power gain. The 1:56 power dividers are driven by eight modules of the same design used in the array face. It may be advantageous to drive the group of eight from a single module via a 1:8 power divider and follow the group of eight by an 8:1 power combiner, which then feeds a 3 dB divider. This arrangement has a few more components than the obvious alternatives, but the additional components are all passive. The added complexity is offset by the inherent fault tolerance of the pool of eight driver modules. When one of these modules fails, the drive power to the array modules is reduced by only one dB.

The total number of active elements in the UCSA design is $5 \times (8 + 112 + 1792) = 9560$ devices. The original PAVE PAWS design employs

$7 \times (1 + 56 + 1792) = 12,943$ devices. As mentioned earlier, the total number of active devices does not necessarily dominate the system design because the cost of the transistors that provide high output power may be a significant cost factor which may be more important in the long run. The design that minimizes the number of active elements is probably the more reliable of the alternatives, and in the future it may also be the lowest in life-cycle cost.

4.5 PARTITIONING OF A CSA INTO MODULES

A module may be defined as a subset of a CSA, which has a single input port and one or more output ports. It is desirable to make all modules identical in order to reduce their manufacturing cost. It may also be desirable to employ higher level modules, composed of a group of lower power modules, with the partitioning determined by cost and the testing approach that minimizes overall cost in both acquisition and maintenance. As a module becomes more complex, the structure that supports the modules becomes simpler. A trade-off of overall cost is the likely result, as illustrated in Figure 4-5.

The partitioning of a transmitter into modules is a design decision that can have a profound effect on acquisition cost as well as the cost of maintaining the system during its service life. It can favor one type of cost over another. If the module were to be defined as a very large component, the cost of keeping spares in the logistics pipeline would be high, while the maintenance system would be made very simple. Ultimately, the whole transmitter could be defined as a replaceable unit. This would be a very nice approach to simplifying the maintenance process, but the cost of the spares would be prohibitive. On the other hand, the lowest replaceable unit could be a single resistor or capacitor, which has a very low probability of failure. The cost of maintaining such a diverse inventory may be a poor choice because most of the components in the pool of spares would be likely to fail because of age rather than use. These systemic level decisions can only be made after a careful analysis of the reliability and ease of maintenance of the system.

It is a good idea to minimize the number of connectors on a module. This reduces the cost of assembly as well as maintenance down time and improves reliability. If a UCSA were to be partitioned into modules, for instance, a 1:4 fan-out ratio design as shown in the PAVE PAWS example, then a 4:1 power combiner could be included so that only a single output connector would be required. For modules that are members of the output rank of a single output port transmitter (not a phased array), these internal power combiners are a subset of the output combiner between the output rank transistors and the transmitter output port. For modules located in the driver level ranks, these internal power combiners are redundant, and they are usually followed by power dividers that feed other driver stages or other members of the output rank. With a small internal combiner ratio, the penalty of this redundancy is offset by the advantages of single RF output connectors per module and interchangeability of all modules in the transmitter. Another advantage of a small internal power

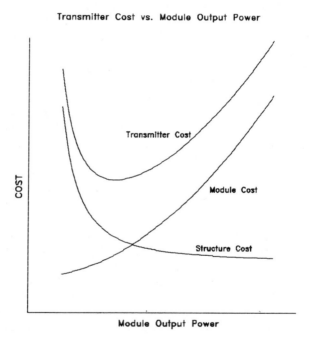

Transmitter Cost vs. Module Output Power

Figure 4-5

combiner ratio is a lower overall output combiner loss. The power combiners that are usually employed internally are either microstrip or compact stripline designs, which are more lossy than larger externally mounted power combiners.

The number of redundant internal power combiner/external power divider pairs in the driver ranks is usually small because the module gain is high. The overhead ratio for the simple PAVE PAWS UCSA example is found by using (4-3) as $O = (1 - 1/15)/15 = 0.062$. Thus, 6.2 percent of the modules have redundant power combiners, a small price to pay for the elimination of multiple output connectors and the advantage of module interchangeability.

4.6 POWER OUTPUT AND RISE TIME CONTROL

There are at least two obvious techniques that can be used to dynamically vary the RF pulse amplitude of a CSA transmitter. The symmetry of the corporate structure can be exploited to partition the transmitter into two identical halves, which have their power output combined in a single large waveguide *magic tee* power combiner. The drive signals going into each half-transmitter are then phase shifted to control the output amplitude. The output of a single output port CSA transmitter also can be varied by switching off modules in the output rank. The switch may be located at the input of each module, and either permits the RF drive signals to reach the module input, or terminates the drive signals in a matched load. Switches of this type are similar to those employed in switched-line phase shifters that employ PIN diodes. The output power varies as the square of the fraction of output rank modules that are turned on. For example, with half of the modules turned off, the output power is reduced to one-quarter of the output produced when all modules are driven. The power output when m out of n modules are energized is $(m/n)^2$ times the power output when all n output rank modules are energized. The difference between the total module power developed mP_m and the transmitter output $(m/n)^2 \times (mP_m)$ is the power dissipated in the terminating loads of the output power combiner. The control of output power in this manner is similar to the action of a *digital-to-analog converter* frequently employed in computer-controlled servo systems. The analogy is exact when the fan-out ratio of the CSA is binary and a binary control switching logic is applied to the successive ranks of the CSA to switch off output rank devices in steps of 2^n. This method could be used to amplitude modulate a CSA from a computer-driven switching network in the driver stages of a CSA. Rise time as well as the amplitude of a train of RF output pulses could be controlled in this manner. The relationship between the output power, power dissipated in the output power combiner loads, and the active fraction of the output rank is shown in Figure 4-6. This also applies to the analysis of the effects of failures on the output power of the transmitter.

When the transmitter is partitioned into two equal sections and phase control of the drive to each is used to control the output amplitude, the following relations apply to the final power combiner in the transmitter:

$$P_o = \frac{P_1 + P_2}{2} + \sqrt{P_1 P_2} \cos\theta \qquad (4\text{-}6)$$

$$P_d = \frac{P_1 + P_2}{2} - \sqrt{P_1 P_2} \cos\theta \qquad (4\text{-}7)$$

P_o is the transmitter output. It is maximum when $\cos\theta$ is equal to 1.0 and the two half-transmitter components, P_1 and P_2, are equal. P_o is equal to $P_1 + P_2$ while the power dissipated in the dummy load terminating the output hybrid, P_d, is zero. When θ is set to 90 degrees, the output is reduced to zero and all of the transmitter power is sent to the dummy load. This is a disadvantage of the phase control scheme because it wastes the power that is not sent to the output port. A second drawback lies in the degree of precision required to reduce the output to a very low level. A phase error of as little as 5 degrees in a 90-degree setting of θ gives a power output that is reduced by 10.6 dB from the full power setting, rather than the intended zero output level. Both efficiency and precision of output amplitude control favor the switched drive technique over the phase control of two half-transmitters in order to vary the amplitude.

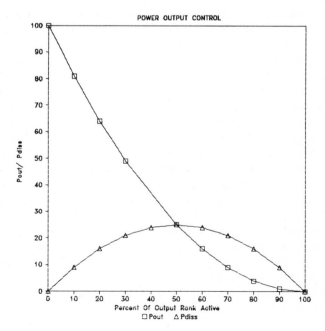

Figure 4-6

Chapter 5

Power Combiner Design

A solid-state transmitter has three major microwave components: the solid-state power amplifiers, the input power divider that delivers RF drive power to the amplifiers, and the output power combiner. This chapter discusses some of the problems involved in designing power combiners. The same conditions apply in general to input power dividers except that insertion loss and power handling are not as critical.

The *power combiner* gathers the output power of all the individual power amplifiers and delivers it to a single power output terminal. The combining is done as a coherent power adder, and the output is the arithmetic sum, less internal losses, of all the output amplifier module powers. It is the ability to add, or combine, the power of many amplifiers with good efficiency that makes solid-state transmitters practical.

An important feature of that efficiency is the ability to provide "graceful degradation," which is a term used to describe a slow falling off in output power as a direct result of individual output power amplifier failure. Because power amplifiers do not last forever, the power loss from a few modules should be planned for, and should result in the least possible loss at the transmitter output.

With many power amplifiers to interconnect, the physical shape and configuration of the combiner as well as the grouping and packaging of the power amplifiers become major design parameters. Most power amplifier modules can be grouped in rather flexible configurations that will meet system designated requirements for replacement, cooling, and prime power connections. However, finding a mechanical space relationship that permits direct, equal length connections to a low-loss RF power combiner greatly restricts that flexibility.

In most cases, the overall driving force in the transmitter design will be the RF interconnections to the combiner. For anyone who has attempted to design a solid-state transmitter, it becomes immediately apparent that module stacking, cooling, and RF interconnections are very interactive. The choices made in selecting and designing the power combiner generally will control the final packaging of the transmitter. Therefore, before proceeding very far, the effect of these choices must be carefully evaluated.

5.1 POWER COMBINER TYPES

Power combiners can be separated into two general types, reactive and hybrid. The hybrid type has resistive terminations at the combining junctions, and under normal combiner operation, power is directed away from the terminations. However, when operated with an imbalance of input power, some of the out of balance power is directed into the terminations. This results in a low standing wave ratio (SWR) at the combiner inputs for all levels of input power. As a result, the combiner performance will not be dependent on input power levels.

With reactive combiners, any imbalance in input power produces reflections from combiner junctions. If the phase of the reflections is carefully controlled, the standing wave ratios may not exceed 2.0 to 1, and this may be acceptable for the power amplifiers[1]. However, some frequency-sensitive phase and amplitude ripples will result. This, and the possible effect on power amplifier life, should be carefully examined when such a design is considered.

The combiner may be binary in form, using two-to-one junctions in cascade. In this case, the number of inputs must be equal to a power of 2, i.e., 2,4,8,16,32, *et cetera*. For other combinations a non-binary 3-, 5-, 6-, or 7-to-1 junction is also used somewhere in the combiner. Because a combiner is the product of cascaded junctions, a 160–amplifier combiner can be realized using a 4, 5, 2, 2, and a 2 combination as a cascaded sequence. Here there would be 40 4:1 combiners, then eight 5:1 combiners, four 2:1 combiners, two 2:1 combiners, followed by one 2:1 combiner as the final power combiner.

As a product sequence, the location of each type of combiner need not be fixed in the sequence. The object of the design is to match power-handling requirements, reduce insertion loss, minimize the number of combiners, and at the same time accommodate a reasonable mechanical configuration for the transmitter.

One further objective is to find a combiner unit suitable as a common element for the input drive side of the power amplifiers. As in the output side, a power combiner is required to feed the amplifier inputs. Here the combiner is operated as a divider with much less power involved.

Reactive and semi-reactive combiners have been designed as single-step combiners. Here the combiner is configured in circular form with the output at the center and the power amplifier connectors spread around the outside[1,2]. While producing a compact design, the transmitter is forced into a round shape where access around the circumference is essential for replacement and servicing of the power amplifiers.

5.2 COMBINER REQUIREMENTS

There are several basic requirements that a power combining network must meet:

(a) The combiner should have low RF insertion loss so that the output power of the transmitter is not wasted in the combining process.

(b) The combiner should have sufficient RF isolation between input connections so that interaction between power amplifiers is negligible.

(c) The combiner should not modify the characteristics of the power amplifiers, including such properties as phase and amplitude response.

(d) The reliability of the combiner should far exceed that of the other transmitter components.

(e) The combiner should exhibit a "graceful degradation" feature and the ability to remain on-line during replacement of power amplifiers. Power dissipation throughout the combiner should be sufficient to accommodate any power amplifier failure scenario.

(f) The mechanical packaging of the combiner should accommodate the requirements of a good power amplifier configuration as well as short, direct, equal phase, low loss interconnections between the amplifiers and the combiner.

5.3 COMBINER SELECTION

With the sizing of the transmitter, the number of power amplifiers is known along with the power levels, so it is possible to quickly run through the power levels that would be encountered for various combining scenarios. Next, consider the power dissipated in each combiner isolation load for various forms of power amplifier failure. It should then be possible to separate the junctions into power-level domains that coincide with the common types of RF connectors and transmission lines. Each domain may then be grouped into separate combiner subassemblies to be used as building blocks of the overall power combiner. As a guide in selecting power handling boundaries, consider the list in Table 5-1 of conservatively rated coaxial lines operating at 1 GHz. The size of an equivalent strip line can be determined by setting the ground plane spacing at roughly the same as the coaxial outer conductor diameter.

Table 5-1

Coaxial Transmission Line Parameters at 1 GHz and Air Coaxial
Losses for 6061–T6 Aluminum

Coax Size	Peak Power	Avg. Power	Loss/100 ft
RG 142B/U	—	90W	4.0 dB
¼ FSJ1(foam)	5kW	300W	4.5 dB
¼ FSJ4(foam)	19kW	800W	3.2 dB
Type N	5kW		
½ in air coax	—	1000W	1.8 dB
⅞ in air coax	61kW	3000W	0.9 dB
1–⅝ in air coax	234kW	6000W	0.45 dB
3–⅛ in air coax	920kW	18kW	0.25 dB

Two things can be quickly obtained from this exercise:

First, we have the difficulty encountered as a physical packaging problem in making connections to each amplifier that will be equal in phase, short, and low loss, and at the same time will conform to the overall transmitter stacking and packaging requirement for the power amplifiers. (When possible, the ideal is to build the first combiner assembly with connectors that interface directly with the power amplifiers.)

Second, we see the way insertion loss quickly adds up. Because insertion loss is highest in the smaller transmission lines, and the distances between input ports are greatest due to the stacking of the power amplifiers, the input portion of the combiner will exhibit the highest loss. A good solution is not to resort to cabling, but instead attempt to include the interconnecting lines in the first stripline assembly, using the lower loss of the stripline to spread out the first combiner group across the amplifier stacking. A four- or eight-to-one combiner assembly is convenient for this purpose.

Frequency bandwidth, in conjunction with input voltage standing wave ratio (VSWR) and insertion loss, must also be considered before proceeding further. The majority of simple combiner junctions employ one-quarter wavelength matching structures. These provide low loss and low standing wave ratios over at least a 10 percent bandwidth. Bandwidths from 15 percent to 20 percent can be achieved with this approach if a doubling of insertion loss is acceptable. The voltage standing wave ratio will probably fall in the 1.2:1 range at band edge *versus* a usually achievable 1.1:1 for the narrower bandwidth. Bandwidth extension may be accomplished by the use of quarterwave transformers in the junction design, and also by cascading in line sections of the divider. This method will result in increased insertion loss[3].

The most common combiners, which have been used in planar stripline construction, are the two-port equal split in-line Wilkinson[4] design, the unequal split in-line Parad and Moynihan[5] design, and the multisection equal split in-line Cohen[6] design, all of which are described in detail for easy design application in Howe[3].

All of the above designs use a terminating resistor imbedded into the circuit so that obtaining high peak and average power dissipation is difficult. In 1975 Gysel[7] described a divider/combiner, which could be realized in planar configuration in which the terminating loads were transformed to a position external to the network where they could be made as large as required and cooled conven ntly. Thus, a design approach is available where power handling is limited only by the transmission line selected. As with the Wilkinson design, the number of combining branches is selective.

After selecting the first combiner assembly, the grouping of the remaining combinations can be made on the basis of power handling, mechanical packaging, and load requirements. It is important to select a configuration with all necessary mechanical interconnections, as well as the lowest loss combiner transmission lines practical, so that the RF insertion loss is as low as possible over the bandwidth of interest. The sum of the power out of all the power amplifiers is achieved at great expense, and the power lost in the power combiner may make the difference between an acceptable transmitter efficiency and one that falls short.

Having selected a first-cut combiner configuration, a study should now be made of the effect of power amplifier failure. With the deployment of hybrid-type combiners, the failure of a series of power amplifiers results in a wave of power that moves through the combiner, loading down the terminating resistors. When one power input to a two-to-one combiner is lost, one-half of the remaining input power is dissipated in the terminating load. The remainder is then passed on to the output. This also means that one-fourth the expected power is passed on to the next hybrid input, which results in a power loss into the terminating resistor of that junction.

Because the power at the second combining level is greater, the power passed into the termination resistor is also greater. This effect continues throughout the combiner until the power inputs for succeeding junctions are closer to a balanced condition where no power is delivered to the terminating resistors.

By choosing various power amplifier failure combinations, the highest power handling requirement can be determined for all the terminating resistors in the power combiner. The terminating resistors should be designed to handle the worst power distribution conditions on a steady-state basis. This will eliminate the failure of terminating resistors that can produce high standing wave ratios at the combiner inputs, which will normally threaten power amplifier stability and usable lifetime.

5.4 CHOICE FACTORS WITH REGARD TO COMBINER TYPES

Regardless of the type of combiner used in the design of the overall power combiner, certain conditions must be satisfied or understood. Some of them are:

(a) Output voltage standing wave ratio:

The power combiner stands between the power amplifiers and whatever load to which the transmitter is delivering power. If as in most radar installations there is a long transmission-line run to the radar antenna, the voltage standing wave ratio as seen by the transmitter will not be low. Careful examination of the power combiner will show that this standing wave ratio will be transferred through the power combiner, and becomes the output standing wave ratio load for each power amplifier in addition to the internal standing wave produced by the combiner itself. The solution is to provide a single high-power circulator/isolator at the combiner output, or a circulator/isolator at the output of each and every power amplifier. The latter solution is usually superior because interaction between power amplifiers is further reduced. Also, a heat sink is usually available, located at the amplifiers, into which the reflected power can be absorbed.

(b) Combiner, power amplifier interaction:

Power transistors, which are the usual devices used in power amplifiers, are operated in distinctly nonlinear modes. When coupled to the power combiner, spurious signals and oscillations may occur, which can be destabilizing to the amplifier and sometimes life-threatening to the output elements. There are several specific effects to be aware of:

(1) The gain bandwidth of the amplifier may be much wider than that of the power combiner and the out-of-band mismatch may present unstable loads to the transistors;

(2) Low-frequency isolation between amplifiers may not exist in the combiner circuitry, resulting in low-frequency oscillations;

(3) The output of harmonic power may be enhanced due to the impedance loading of the combiner at certain out-of-band frequencies. The production of harmonic power may require the introduction of RF filtering, so that the transmitter may conform to the spectrum signature requirements imposed on the end equipment. One other advantage in choosing to locate a circulator-isolator at each power amplifier output is that it can also be utilized in minimizing the harmonics if the reverse loss bandwidth can be made to extend much wider than the operational bandwidth.

(c) Phase and amplitude ripple effects:

Phase and amplitude variations as a function of frequency are a problem when designing large combiner networks. The ripples arise as the result of combiner junction discontinuities separated by long electrical distances, which increase as the bandwidth requirements increase. As a function of frequency, the reflections from the discontinuities add and cancel in a cyclical manner. The result is phase and amplitude variation across the frequency band.

The voltage standing wave ratio of each combiner element should be carefully controlled with design goals for hybrid designs set in the range of 1.1 to 1 to 1.15 to 1. With reactive combiners, goals of 1.1 to 1 or lower should be set because this type of junction can be unbalanced easily.

In calculating the phase and amplitude effects of reactive combiners, we should note that the branch arms of a two-to-one combiner rapidly increase to the 2.0:1 standing wave ratio level with unbalanced input power resulting in much larger phase and amplitude errors than with hybrid combiners. Hybrid combiners remain constant in standing wave ratio independent of input power balance. Flat bandwidth response performance for cascaded reactive combiners is very difficult to achieve. To reduce this effect, reactive combiners are sometimes surrounded by hybrid combiners. Occasionally, the reactive combiner is used to accomplish the combining process all in one step.

5.5 POWER COMBINER DESIGNS

The coaxial N-way hybrid power divider, described by Wilkinson[4], was introduced in 1960, which permitted in-phase, equal-power power dividers to be realized in other than binary configurations. However, the most popular version of this hybrid is the two-to-one combiner, which can be fabricated in planar form. Figure 5-1(a) is a schematic of the divider. The 70.7–ohm lines are one-quarter wavelength long at center frequency. A compensated version for wider bandwidth use is shown in Figure 5-1(b). Here, a one-quarter wavelength matching transformer has been added at the input port.

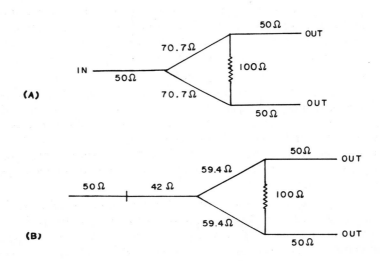

Figure 5-1

The most restrictive design requirement in the divider is imposed on the 100–ohm resistor, which must be small enough not to disturb the RF circuit performance. This can be quite difficult for high power designs where the resistor must hold off high voltage while at the same time conducting high average power into a cooling heat sink.

In 1963, Parad and Moynihan[5] introduced a two-to-one in-phase hybrid divider, which offered arbitrary power division. The circuit is shown in Figure 5-2 along with design formulas.

$$\frac{\text{Power at Port 3}}{\text{Power at Port 2}} = K^2 \tag{5-1a}$$

$$R = 50 \left(\frac{1+K^2}{K}\right) \tag{5-1b}$$

$$Z_1 = \left(\frac{K}{1-K^2}\right)^{1/4} 50 \tag{5-1c}$$

$$Z_2 = K^{3/4}(1+K)^{1/4} \tag{5-1d}$$

$$Z_3 = \frac{(1+K^2)^{1/4}}{K^{5/4}} \tag{5-1e}$$

$$Z_4 = 50\sqrt{R} \tag{5-1f}$$

$$Z_5 = \frac{50}{\sqrt{R}} \tag{5-1g}$$

In 1975, Gysel[7] extended the N-way power divider design by providing a planar configuration with a quarter-wavelength transformation, which places the RF loads outside of the circuit. Figure 5-3 is a schematic of the circuit.

Closed solutions for optimum design do not exist. A computer-aided design method is included in the appendix for determining VSWR, insertion phase, and insertion loss.

In 1968, Cohen[6] increased the bandwidth of the in-phase, equal-power divider by cascading a number of quarterwave sections with resistive terminations at the end of each section. Howe[3] described the design along with the applicable design tables and curves.

In 1980, Sanders[2] reported the development of a radial line combiner design that can accommodate approximately 10 to 200 inputs in a one-step approach. In a 110-branch model, isolation between inputs was reported to be at least 18 dB, the input VSWR below 1.2 to 1, and the insertion loss was 0.25 dB maximum over a 15.4 percent band.

Quadrature hybrids can be of two forms, branched and parallel coupled. Figure 5-4 is a two-branched 90-degree hybrid, and Figure 5-5 is a parallel coupled hybrid.

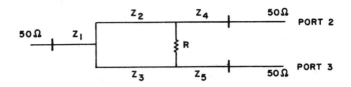

Z_1 THRU Z_5 ARE ONE QUARTER WAVELENGTH LONG AT
CENTER FREQUENCY

Figure 5-2

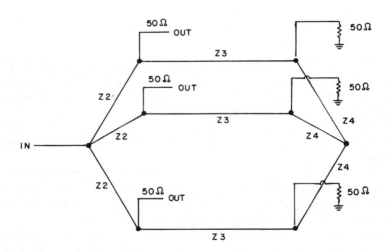

ALL TRANSFORMERS ARE ONE QUARTER WAVELENGTH
AT CENTER FREQUENCY

Figure 5-3

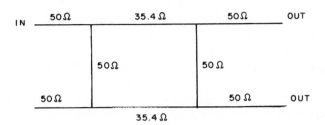

THE FOUR ARMS OF THE BRANCH ARE
ONE QUARTER WAVELENGTH AT CENTER FREQUENCY

BRANCH COUPLER

Figure 5-4

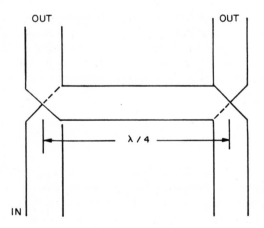

PARALLEL COUPLER

Figure 5-5

For 3 dB coupling in a 50–ohm medium, even-mode impedance is 100 ohms, and odd-mode impedance is 25 ohms for the parallel coupled design.

The bandwidth of both designs may be extended by the use of additional sections. Howe[3] is a design source for this approach.

Both the branch and parallel designs can be made to handle large peak and average power in air stripline or large coaxial lines. The parallel branch, due to its relatively small size, is quite attractive as a final, or nearly final, combiner in a high-power system of the 250kW peak, 10kW average size operating at 1GHz or lower. The required 90-degree input phasing requirement may be accommodated by placing a low-power quadrature hybrid at a symmetrical location in the input power divider network that feeds the power amplifiers.

5.6 POWER DISTRIBUTION CALCULATIONS

With the loss of power amplifier modules at the input to a power combiner, there is a loss in output power and, if the combiner is of the hybrid design, a loss of power into the terminating resistors within the combiner. In a two-to-one hybrid combiner, the loss of input power at one input results in one-half of the input power being directed into the terminating resistor, while the remaining half reaches the output port. The same relationship holds for a hybrid combiner consisting of many input ports. The equation

$$K = \left(\frac{M}{N}\right)^2 \tag{5-2}$$

where K is the decrease in output power, M is the full number of inputs, and N is the number of operating inputs, can be used to determine the loss in output power for a loss in input modules. The difference between the output power and the input power for N operating amplifiers is the power delivered into the terminating resistors within the power combiner. This relationship can be used to estimate the "graceful degradation" characteristic of the transmitter as well as show the worst-case power dissipation. This occurs when one-half of the modules are down and simultaneously one-fourth of the normal full power of the transmitter is being delivered to the power combiner's internal terminating resistors.

In order to determine the individual combiner dissipation power levels required in the terminating resistors, the following relationships can be used to trace the power distribution from combiner to combiner:

$$P_{out} = \frac{P_1 + P_2}{2} + \sqrt{P_1 P_2}\cos\theta \qquad\qquad (5\text{-}3a)$$

$$P_3 = P_1 + P_2 - P_{out} \qquad\qquad (5\text{-}3b)$$

P_1 and P_2 are the incident powers of the two-to-one combiner. θ is the phase angle between P_1 and P_2, P_3 is the power delivered to the terminating resistor. For quadrature hybrids θ is the phase deviation from 90 degrees.

For a reliable design to be achieved, the power handling terminating resistors within the combiner should be designed to survive steady-state operation at these power levels.

For N-way combiners, power dissipation in the terminating resistors follows the relationship (5-2); however, the load distribution will not be equal between resistors. A reasonable assumption is that the lost power dissipates evenly between the resistors of the inactive ports directly adjacent to the active input ports.

5.7 POWER HANDLING

The power handling capability of the power combiner should be one of the first considerations when undertaking a solid-state transmitter design. With so many components to interconnect, it is a major task to evaluate and select the proper combiners to create an efficient, low loss, high-power handling power combiner that will meet all of the predefined customer and system requirements. The first task when looking at power handling is to establish the power handling potential of the transmission lines that are under consideration.

There are two aspects of power, peak and average. Peak power handling is based on the maximum voltage gradient produced for a given input power. When that power results in a maximum voltage gradient, which is judged likely to produce voltage breakdown, then this is the maximum peak power handling capability of the component or transmission line. Average power handling is based on temperature rise and is directly related to insertion loss and thermal cooling.

Peak and average power notation is common in pulse systems, where the maximum amplitude of a short pulse is the peak power. Average power is a pulse train converted arithmetically to a continuous wave. When pulses are short, there is usually a large ratio between peak and average power. For long pulses, where this ratio is much less, it is usually necessary to assume the long pulse to be the average as well as the peak

156

power. When there is enough average power to elevate the temperature of the component, or transmission line, the peak power handling capability must be revaluated. This is because the voltage gradient breakdown level is an inverse function of temperature.

5.7.1 Power Handling and Voltage Gradient

When calculating peak power handling, what voltage gradient value should a designer employ? It is common practice to use 30,000 volts per centimeter as the breakdown voltage gradient for microwave components. This is for sea level (14.7 pounds per square inch absolute air pressure) air pressure, and 70-degree Fahrenheit temperature (so-called room temperature). This is an optimistic value to use as it is the published breakdown in statics between smooth, polished, spherical surfaces. Another value 'quoted is 10,000 volts per centimeter, the breakdown for sharp, polished needle points. This is somewhat closer to the realistic value for surfaces sometimes encountered in microwave devices, but it is quite pessimistic. A good compromise that is in agreement with realistic historical experience is 15,000 volts per centimeter, and this is the recommended value for peak power calculations.

When calculating peak power, it is usually assumed that standard atmospheric conditions exist and the usual recommended value for voltage breakdown is used; that is, the air pressure is at 14.7 pounds per square inch absolute (sea level), and the temperature is 70 degrees Fahrenheit (room temperature). However, voltage breakdown varies linearly with both absolute temperature and absolute pressure over a reasonable range. When the air pressure is increased, the voltage breakdown level increases. When the temperature is increased, the voltage breakdown level decreases.

Because peak power varies as the voltage gradient squared, the effects of temperature and air pressure change can be conveniently calculated as ratios. Increasing the air pressure from 14.7 pounds per square inch absolute (PSIA) to 29.4 PSIA, doubles the voltage breakdown level and increases the peak power handling four times. Increasing the temperature from 70 degrees Fahrenheit (°F) (290 Kelvins) to 271° F (406 Kelvins) decreases the voltage breakdown level by 1.4 and decreases the peak power handling by a factor of two. The air pressure effect is linear for about three atmospheres. We must be careful about temperature increases because they are not distributed uniformly as in the case of air pressure.

5.7.2 Peak Power Handling in the Presence of Small Voids

Another important factor in high-power design is the part that very small voids, or gaps, play when allowed to exist within dielectrically loaded components.

Dielectric loading is often used to increase power handling. It is assumed that the dielectric material completely fills the volume. This assumption may be true, but if small voids or gaps exist between dielectric sections, or between the dielectric and the conductor walls, then the risk of power breakdown becomes high.

Across very small air gaps, the electrical potential D is continuous. That is, D, which equals the electric field strength times the dielectric constant, cannot equal just the electric field. The result is an electric potential build-up within the void equal to the relative dielectric constant times the electric field. This is why dielectrically loaded striplines often fail at the point where the fit between parts is not the best. This is also why it is difficult to maintain the rated power handling equal to that of a cable in the connector interface area.

When calculating the peak power handling of components, the field strength in the sensitive areas should be assumed to be higher by a factor equal to the dielectric constant. This is a valid argument against the use of very high dielectric constant material in high-power applications.

5.8 COAXIAL LINES OPERATING IN THE TEM MODE

5.8.1 Peak Power

The peak power handling of a coaxial line is

$$P = (0.674) E^2 \sqrt{\epsilon} \log_{10} \left(\frac{b}{a}\right)^2 \qquad (5\text{-}4)$$

where P = peak power (W), E = voltage gradient, t the surface of the center conductor (V per cm), ϵ = relative dielectric constant, a = radius of inner conductor (inches), and b = radius of the outer conductor (inches).

For a dielectrically loaded coaxial line, the electric field strength at the center conductor in a small void in the dielectric is

$$E_a = 215.5 \sqrt{\frac{\epsilon}{a}} \sqrt{\frac{P}{Z_0}} \qquad (5\text{-}5)$$

where E_a = electric field strength in the void at the center conductor (V per cm), ϵ = relative dielectric constant (between the conductors), a = center conductor radius (inches), P = peak power (W), and Z_0 = line impedance (ohms).

5.8.2 Line Impedance

The impedance of a coaxial line is

$$Z_0 = \frac{138\log}{\sqrt{\epsilon}} \ 10 \ \frac{b}{a} \tag{5-6}$$

where Z_0 = impedance (ohms), ϵ = relative dielectric constant between conductors, a = center conductor radius (inches), and b = outer conductor radius (inches).

5.8.3 Insertion Loss

The insertion loss of an air-filled copper coaxial line is

$$\alpha = 2.98 \ (10^{-9})(\sqrt{f}) \ \frac{1}{b} \left(1+\frac{b}{a}\right)\left(\frac{1}{\ln\left(\frac{b}{a}\right)}\right) \tag{5-7}$$

where α = insertion loss (dB/inch), f = frequency (Hz), a = center conductor radius (inches), and b = outer conductor radius (inches).

5.8.4 Average Power Handling

The average power handling of a coaxial line is strictly an operating temperature limitation. The average power flowing in the line produces heat, and following an initial warm-up period, stabilized elevated temperatures occur at the outer and inner conductor surfaces, dependent on heat loss paths.

A common standard is to measure the amount of average power it takes to produce a 40-degree Celsius temperature differential between inner and outer conductors, and to define this differential as the average power handling capability. The maximum environmental temperature must be added to this. Whether this rating is acceptable depends on other factors. Having arrived at the expected temperature due to the average power dissipation in the line, the peak power handling must be reviewed and corrected for the temperature increase above the original value of the room temperature voltage gradient breakdown. In a 50-ohm coaxial line, 70 percent of the insertion loss occurs in the center conductor. A good approximation of the temperature rise for air-filled lines can be obtained by assuming that all the heat is lost by thermal radiation. This is reasonable because the thermal conduction losses along the center conductor, the only direct conduction path, are high and will not contribute much to heat loss.

After calculating the power loss per unit length from the insertion loss relationship and the expected average power, the temperature rise is found for the surface area of the unit length by using the thermal radiation equation:

$$P = \epsilon_t \sigma \, (T^4 - T_0^4) \tag{5-8}$$

where P = radiated power (W/cm^2), ϵ_t = total thermal emissivity of the surface, σ = Stefan-Boltzmann constant (5.67) $\times$ 10^{-12} W$\times$cm$^{-2}\times$K^{-4}), T = temperature of radiating surface in Kelvins, and T_0 = temperature of the surroundings in Kelvins. The emissivity of copper is approximately 0.1, and that of a black surface is 1.0. Normal ambient temperature is 290 K, and 0 degrees Celsius is 273 K.

In cases where the thermal problem is more complicated, a detailed thermal analysis should be performed.

5.9 STRIPLINE OPRATING IN THE TEM MODE

Stripline consists of a metal strip suspended equidistant between two ground planes. The space between the ground planes may be air or filled with dielectric.

5.9.1 Peak Power

Peak power handling in stripline is approximately equivalent to the peak power handling of the coaxial line interfacing it, if the outer conductor diameter matches the ground plane spacing and is of about the same impedance. This is true even if both are dielectrically loaded (equal dielectric constants). In air stripline, the peak power handling can be increased by rounding the corners of the center conductor. If a fair amount of peak power is being considered, the safety factor should be verified by high-power testing.

5.9.2 Line Impedance

Howe[3, Ch.2] is a good source for various solutions used to determine the characteristic impedance of stripline. Howe includes the method developed by Rosenzweig for the offset center conductor lines used in some three-layer stripline construction. Also, for calculations regarding center conductor thickness, one-ounce clad copper is 0.0014 inches thick, two-ounce copper is 0.0028 inches thick, *et cetera*.

5.9.3 Insertion Loss

Howe[3, Ch.1] also presents tables and formula for stripline insertion loss using Cohen, Rosenzweig, and Getsinger as sources. The conductor losses are based on copper conductors. If the conducting surface is a metal other than copper, the results should be scaled by the ratio of the surface resistivity (Ω^2) of this metal to that of copper.

5.9.4 Average Power Handling

For air stripline, the average power handling can be estimated using the method described for coaxial lines. For dielectrically loaded lines, Howe[3, Ch.1] lists data for calculating the temperature rise of the center conductors for several dielectric materials, for ground plane spacings between 0.062 and 0.125 inches, and for frequencies to 10 GHz.

5.10 MICROSTRIP LINE OPERATING IN THE DOMINANT MODE

A microstrip transmission line consists of a metalized strip and a ground plane separated by a dielectric.

5.10.1 Peak Power

Peak power handling to 10kW at 9 GHz for a 50 ohm 1/16–inch Teflon-fiberglass microstrip line has been reported under matched conditions. A reliable method for calculating peak power handling is not available.

5.10.2 Line Impedance

White[8] has collected work by early authors and summarized the characteristic impedance relationships for narrow and wide strip geometries as well as listing a Fortran program for their calculation.

5.10.3 Insertion Loss

The insertion loss is calculated in two parts, conductor loss and dielectric loss. The conductor loss for one ounce copper strip in dB/ft is

$$\alpha_c = 7.25 \times 10^{-5} \left(\frac{1}{h}\right) (f\epsilon)^{1/2} \tag{5-9}$$

where h = thickness of dielectric (inches), f = frequency (MHz), and ϵ = dielectric constant (relative to air). The dielectric loss in dB/ft is

$$\alpha_d = 2.78 \ (10^{-2}) \ fF_P \ \sqrt{\epsilon} \tag{5-10}$$

where f = frequency (MHz), F_p = dielectric loss tangent, and ϵ = dielectric constant (relative to air).

5.10.4 Average Power

A temperature rise of 50 degrees Celsius has been measured above 20 degrees ambient temperature for a 7/16–inch wide, 1/16–inch thick Teflon-impregnated fiberglass baseline carrying 300 watts CW at 3 GHz.

5.11 RECTANGULAR WAVEGUIDE OPERATING IN THE TE$_{10}$ MODE

Rectangular waveguide discussed here is operated with one mode, the dominant TE$_{10}$ mode, where the a dimension is the inside wide dimension and b is the inside narrow dimension. Two metals in common use are 6061–T6 aluminum, and oxygen-free copper.

5.11.1 Peak Power

Peak power handling, for a given electric field gradient, for a rectangular waveguide is

$$P = E_m^2 \ (6.63)(10^{-4})(ab) \ \sqrt{1 + \left(\frac{\lambda}{2a}\right)^2} \tag{5-11}$$

where P = peak power (W), E_m = maximum electric field gradient (V/cm), a = inside wide dimension of the waveguide (inches), b = inside narrow dimension of the waveguide (inches), and λ = free-space wavelength (inches).

5.11.2 Insertion Loss

The insertion loss for air-filled oxygen-free copper waveguide is:

$$a_c = \frac{1.107}{a^{3/2}} \left[\frac{\frac{1}{2} \frac{a}{b} \left(\frac{f}{f_c}\right)^{3/2} + \left(\frac{f}{f_c}\right)^{-1/2}}{\sqrt{\left(\frac{f}{f_c}\right)^2 - 1}} \right] \qquad (5\text{-}12)$$

where a_c is (dB/100 ft), a = inside wide dimension of waveguide (inches), b = inside narrow dimension of waveguide (inches), f = frequency, and f_c = the cut-off frequency, which is found by dividing the speed of light (c) by twice the a dimension. The insertion loss of 1100–type aluminum is 1.28 times the loss of copper, and the loss for 6061–T6 aluminum is 1.53 times the copper loss values. Most aluminum waveguides are 6061 aluminum and not 1100 aluminum, which is much too soft structurally.

5.11.3 Average Power

Again, average power handling is based on temperature rise and the determination of how hot is too hot. Waveguides, unlike coaxial lines, can be conveniently liquid-cooled to maintain an acceptable operating temperature. Natural cooling by radiation can be calculated using (5-8). If this is not sufficient, and water cooling can be used, the relationship is

$$P = 264 \, Q_w \, (T_2 - T_1) \qquad (5\text{-}13)$$

where P = heat removed (W), Q_w = water flow (gallons/minute), and T_1 and T_2 are the inlet and outlet water temperatures respectively (Kelvins).

5.12 RIDGED WAVEGUIDE OPERATING IN THE TE$_{10}$ MODE

There are conditions where, in order to obtain wide bandwidth and to avoid coaxial operation, ridged waveguides may be attractive. Data on double-ridged waveguides with a bandwidth ratio of 2.4 to 1 is listed.

5.12.1 Peak Power

For double-ridged waveguide with a bandwidth ratio of 2.4:1, the peak power handling, based on 15,000 V/cm maximum field gradient is:

Waveguide	Frequency Range(MHz)	Peak Power Handling(MW)
D2967	175— 420	62.0
D1943	267— 640	26.5
D1233	420—1000	10.7
D810	640—1530	4.6
D617	840—2000	2.6
D346	1500—3600	0.84

5.12.2 Insertion Loss

The insertion loss of an oxygen-free copper double-ridged waveguide with a 2.4:1 bandwidth ratio is:

Waveguide	Frequency Range(MHz)	Attenuation(dB/ft)
D2967	157— 420	0.00023
D1943	267— 640	0.00043
D1233	420—1000	0.00085
D810	640—1530	0.0016
D617	840—2000	0.0024
D346	1500—3600	0.0058

5.12.3 Average Power

Average power handling capability is based on acceptable temperature rise. The method used to determine this is described under the various headings for average power listed for the other transmission lines.

5.13 COMPONENT ANALYSIS USING ABCD MATRICES

5.13.1 Introduction

Microwave circuit analysis can be readily applied to component analysis using ABCD matrices and a digital computer. Ritchards[9], White[9], and Gupta[11], to name only a few, have described methods for this approach. The engineer may write his own programs, or use recently developed computer-aided designs (CAD), such as COMPACT or SUPER COMPACT, which are currently on the market. To cover those cases where a CAD program may not be available, or if the designer wishes to become familiar with the details of this method, some of the basics are introduced.

5.13.2 Matrix Networks

The ABCD Matrix
The ABCD matrix represents a two-port network:

$$\begin{pmatrix} V_1 \\ I_1 \end{pmatrix} = \begin{pmatrix} A & B \\ C & D \end{pmatrix} \begin{pmatrix} V_2 \\ I_2 \end{pmatrix}$$

WHERE:

$$V_1 = AV_2 + BI_2$$

$$I_1 = CV_2 + DI_2 \tag{5-14}$$

Because this network is a relationship between voltage and current at the input end, and the voltage and current at the output end, it is suitable for cascade connections. The result obtained from the ABCD matrix networks connected end-to-end is the matrix product:

$$\begin{bmatrix} A & B \\ C & D \end{bmatrix} = \begin{bmatrix} A_1 & B_1 \\ C_1 & D_1 \end{bmatrix} \times \begin{bmatrix} A_2 & B_2 \\ C_2 & D_2 \end{bmatrix} = \begin{bmatrix} A_1A_2+B_1C_2 & A_1B_2+B_1D_2 \\ C_1A_2;D_1C_2 & C_1B_2+D_1D_2 \end{bmatrix} \tag{5-15}$$

If the circuits are reciprocal $AD - BC = 1$.

The following circuit parameters may be obtained from an ABCD matrix.

The voltage transmission coefficient is

$$T = \frac{A + BY_0 + CZ_0 + D}{2} \tag{5-16}$$

where $Y_0 = 1/Z_0$ and Z_0 = the characteristic line impedance, which is usually 50 ohms for a coaxial line, and for the waveguide case is normalized to 1. The insertion loss is

$$\text{Loss(dB)} = 20\log_{10}(\text{Absolute Value of } T) \tag{5-17}$$

The insertion phase of the network is the phase angle of the complex value of T:

$$\text{Insertion Phase} = \tan^{-1}\left(\frac{T_{imaginary}}{T_{real}}\right) \tag{5-18}$$

The input voltage reflection coefficient is

$$\Gamma_{in} \frac{A + BY_0 - CZ_0 - D}{A + BY_0 + CZ_0 + D} \tag{5-19}$$

and the input voltage standing wave ratio VSWR is

$$\text{VSWR}_{in} = \frac{1 + |\Gamma_{in}|}{1 - |\Gamma_{in}|} \tag{5-20}$$

The reflection cofficient angle is the phase angle of the complex value of the voltage reflection coefficient:

$$\text{Reflection Angle} = \tan^{-1}\left(\frac{\Gamma_{imaginary}}{\Gamma_{real}}\right) \tag{5-21}$$

The output voltage reflection coefficient is

$$\Gamma_{out} = \frac{-A + BY_0 - CZ_0 - D}{A + BY_0 + CZ_0 + D} \tag{5-22}$$

and the output VSWR is found from the absolute magnitude of the output reflection coefficient:

$$\text{VSWR}_{out} = \frac{1 + |\Gamma_{out}|}{1 - |\Gamma_{out}|} \tag{5-23}$$

The Y-Matrix
    The Y-matrix as a two-port matrix is

$$\left(\begin{array}{c} I_1 \\ I_2 \end{array}\right) = \left(\begin{array}{cc} Y_{11} & Y_{12} \\ Y_{21} & Y_{22} \end{array}\right) \times \left(\begin{array}{c} V_1 \\ V_2 \end{array}\right)$$

WHERE : $\quad I_1 = Y_{11} V_1 + Y_{12} V_1$

$$I_2 = Y_{21} V_1 + Y_{22} V_2 \tag{5-24}$$

The *Y*-matrix may be used to parallel two two-port networks by adding the *Y*-matrices.

$$\left[\begin{array}{cc} Y_{11} & Y_{12} \\ Y_{21} & Y_{22} \end{array}\right] = \left[\begin{array}{cc} Y_{11}^1 & Y_{12}^1 \\ Y_{21}^1 & Y_{22}^1 \end{array}\right] + \left[\begin{array}{cc} Y_{11}^2 & Y_{12}^2 \\ Y_{21}^2 & Y_{22}^2 \end{array}\right] = \left[\begin{array}{cc} Y_{11}^1 + Y_{11}^2 & Y_{12}^1 + Y_{12}^2 \\ Y_{21}^1 + Y_{21}^2 & Y_{22}^1 + Y_{22}^2 \end{array}\right]$$

$$\tag{5-25}$$

To combine two ABCD matrices, which are connected in parallel, first convert both matrices to *Y*-matrices, add them together, and then convert the sum back to an ABCD matrix:

$$\left[\begin{array}{cc} Y_{11} & Y_{12} \\ Y_{21} & Y_{22} \end{array}\right] = \frac{1}{B} \left[\begin{array}{cc} D & -(AD - BC) \\ -1 & A \end{array}\right]$$

$$\left[\begin{array}{cc} A & B \\ C & D \end{array}\right] = \frac{1}{-Y_{21}} \left[\begin{array}{cc} Y_{22} & 1 \\ (Y_{11} Y_{22} - Y_{12} Y_{21}) & Y_{11} \end{array}\right]$$

$$\tag{5-26}$$

This operation may be combined into one step, where $A_3B_3C_3D_3$ is the resulting ABCD matrix:

$$
\begin{bmatrix} A_3 & B_3 \\ C_3 & D_3 \end{bmatrix} = \begin{bmatrix} \dfrac{D_1}{B_1} + \dfrac{D_2}{B_2} & \overbrace{\dfrac{-(A_1D_1-B_1C_1)}{B_1}}^{} + \overbrace{\dfrac{-(A_2D_2-B_2C_2)}{B_2}}^{} \\[3mm] \dfrac{-1}{B_2} + \dfrac{-1}{B_2} & \dfrac{A_1}{B_1} + \dfrac{A_2}{B_2} \end{bmatrix}
$$

$$
\begin{bmatrix} A_3 & B_3 \\ C_3 & D_3 \end{bmatrix} = \begin{bmatrix} \dfrac{A_1B_2+A_2B_1}{B_1+B_2} & \dfrac{B_1B_2}{B_1+B_2} \\[3mm] C_1+C_2+\dfrac{(A_2-A_1)(D_1-D_2)}{B_1+B_2} & \dfrac{D_1B_2+D_2B_1}{B_1+B_2} \end{bmatrix} \qquad (5\text{-}27)
$$

and A_1, B_1, C_1, and D_1 are elements of the first ABCD matrix, and A_2, B_2, C_2, and D_2 are elements of the second matrix.

The Z-Matrix

The two-port Z-matrix is

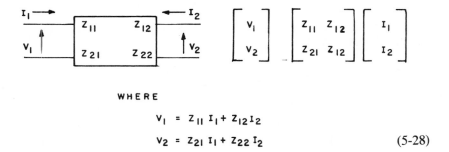

$$
\begin{bmatrix} V_1 \\ V_2 \end{bmatrix} = \begin{bmatrix} Z_{11} & Z_{12} \\ Z_{21} & Z_{12} \end{bmatrix} \begin{bmatrix} I_1 \\ I_2 \end{bmatrix}
$$

WHERE

$$
V_1 = Z_{11}I_1 + Z_{12}I_2
$$
$$
V_2 = Z_{21}I_1 + Z_{22}I_2 \qquad (5\text{-}28)
$$

The Z-matrix may be used to connect two port networks in series by adding the Z matrices:

$$\begin{bmatrix} z_{11} & z_{12} \\ z_{21} & z_{22} \end{bmatrix} = \begin{bmatrix} z_{11}^1 & z_{12}^1 \\ z_{21}^1 & z_{22}^1 \end{bmatrix} + \begin{bmatrix} z_{11}^2 & z_{12}^2 \\ z_{21}^2 & z_{22}^2 \end{bmatrix} = \begin{bmatrix} z_{11}^1 + z_{11}^2 & z_{12}^1 + z_{12}^2 \\ z_{21}^1 + z_{21}^2 & z_{22}^1 + z_{22}^2 \end{bmatrix}$$

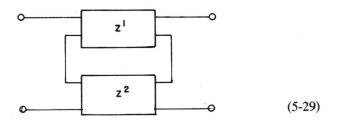

$$(5\text{-}29)$$

To connect two ABCD matrices in series, convert the matrices to Z-matrices, sum them together, and convert back to an ABCD matrix:

$$\begin{bmatrix} Z_{11} & Z_{12} \\ Z_{21} & Z_{22} \end{bmatrix} = \frac{1}{C} \begin{bmatrix} A & (AD - BC) \\ 1 & D \end{bmatrix}$$

$$\begin{bmatrix} A & B \\ C & D \end{bmatrix} = \frac{1}{Z_{21}} \begin{bmatrix} Z_{11} & Z_{11}Z_{22} - Z_{12}Z_{21} \\ 1 & Z_{22} \end{bmatrix} \qquad (5\text{-}30)$$

The operation can be combined into one step, where $A_3B_3C_3D_3$ is the resulting ABCD matrix:

$$\begin{bmatrix} A_3 & B_3 \\ C_3 & D_3 \end{bmatrix} = \begin{bmatrix} A_1A_2 + B_1B_2 & A_1B_2 + B_2D_2 \\ C_1A_1 + D_1C_2 & C_1B_2 + D_1D_2 \end{bmatrix} \qquad (5\text{-}31)$$

and A_1, B_1, C_1, and D_1 are elements of the first ABCD matrix, and A_2, B_2, C_2, and D_2 are elements of the second matrix.

5.13.3 Circuit Representation

Circuit representation of three common two-port circuits and their ABCD matrix elements are

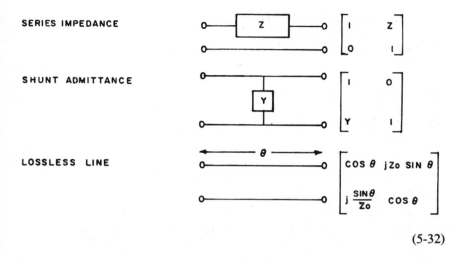

$$(5-32)$$

For the series impedance: $Z = a + jb$, a = resistance in ohms, $b = +\omega L$ or $-\omega c$ in ohms. For the shunt admittance: $Y = c + ja$, c = admittance in mhos, $d = -1/\omega L$ or $+1/\omega c$ in mhos.

5.13.4 Computer Program Language

The computer program language used is Fortran. It is familiar to most engineers, and the programs should be easily adapted to most compiler format requirements provided complex number routines are available.

In order to analyze coaxial or stripline power dividers, four subroutines are employed. The functions performed are:

(1) Form an ABCD matrix of a shunt resistive admittance, or a series resistance in a transmission line of zero electrical length;
(2) Form an ABCD matrix of a transmission line of impedance Z_0 and electrical length θ;
(3) Form the product of two cascaded ABCD matrices, or the sum of the two matrices that are connected in parallel;
(4) Form a new matrix from an old matrix, or transpose the matrix.

The first subroutine, described as (1) above is:

```
SUBROUTINE ABCD(M,Y,IND)
MCOMPLEX M(2,2)
M(1,1)=CMPLX(1.0,0.0)
```

```
      M(2,2)=COMPLEX(1.0,0.0)
      GO TO (1850,1880)IND
1850  M(1.2)=COMPLX(Y,0.0)
      M(2,1)=0.0
      GO TO 1900
1880  M(2,1)=COMPLX(Y,0.0)
      M(1,2)=0.0
1900  RETURN
      END
```

The call from the main program to use this subroutine is:

```
    CALL ABCD(B,YR,2)
```

The subroutine variables, Y and IND, are transmitted from the main program as YR and 2, the ABCD matrix M is returned to the main program as matrix B, a shunt admittance matrix. The shunt admittance Y is real, representing a shunt resistor, such as a terminating load, which is schematically in parallel with the transmission line and is equal in value to $1/R$, where R is the load resistance in ohms. When IND is set equal to 1, the subroutine forms a series impedance ABCD matrix and Y is in ohms, the series resistor.

The next subroutine, (2), which forms a transmission line, is:

```
      SUBROUTINE ABCDXL(M,Z0,L,F)
      REAL L
      COMPLEX J,M(2,2)
      Q=29979.25/2.54
      G=Q/F
      J=CMPLX(0.0,1.0)
      PHI=2.0*3.13415926*(L/G)
      M(1.1)=COMPLX(COS(PHI),0.0)
      M(2,2)=M(1,1)
      M(1,2)=J*Z0*SIN(PHI)
      M(2,1)=J*(1.0/Z0)*SIN(PHI)
      RETURN
      END
```

The call from the main program is:

```
    CALL ABCDXL(A,Z2,L,FREQ)
```

In the subroutine, Z_0, L, and F are transmitted from the main program as Z2, L, and FREQ, and the ABCD matrix M is returned to the main program as matrix A. Z_0 is the line impedance in ohms, L is the physical length of the transmission line in inches (usually one-quarter wavelength at midband). FREQ is frequency in megahertz. G is wavelength in inches, in this case the free-space air wavelength used to form the frequency dependent dimensionless ratio L/G as the electrical length modifier. PHI is converted to radians which is the angle argument in FORTRAN. The wavelength term G would require modification for waveguide or dielectrically loaded transmission lines.

The subroutine to perform function (3), for cascaded or parallel matrices, is:

```
      SUBROUTINE ABCDSP(M1,M2,M3,IND)
      COMPLEX M1(2,2),M2(2,2),M3(2,2),DEN
      GO TO (2250,2300)IND
2250  M3(1,1)=M1(1,1)*M2(1,1)+M1(1,2)*M2(2,1)
      M3(1,2)=M1(1,1)*M2(1,2)+M1(1,2)*M2(2,2)
      M3(2,1)=M1(2,1)*M2(1,1)+M1(2,2)*M2(2,1)
      M3(2,2)=M1(2,1)*M2(1,2)+M1(2,2)*M2(2,2)
      GO TO 2360
2300  DEN=M1(1,2)+M2(1,2)
      M3(1,1)=(M1(1,1)*M2(1,2)+M2(1,1)*M1(1,2))/DEN
      M3(1,2)=m1(1,2)*M1(1,2)/DEN
      M3(2,1)=(M1(1,2)+M2(2,1))+((M2(1,1)-M1(1,1))*
     +(M1(2,2)-M2(2,2)))/DEN
      M3(2,2)=(M1(2,2)*M2(1,2)+M2(2,2)*M1(1,2))/DEN
2360  RETURN
      END
```

The call from the main program is:

```
      CALL ABCDSP(A,B,C,1)
```

Subroutine variables M1, M2, and IND, are transmitted from the main program as A, B, and 1, and the resultant matrix M3 is returned to the main program as matrix C. When IND is 1, as in this example, the two matrices are multiplied from left to right, with M1 being on the left, to form matrix M3. If IND is set at 2 then M1 and M2 are added in parallel to form matrix M3 using the Y-matrix relationships.

The last subroutine (4), used to form a new matrix or transpose, is:

```
      SUBROUTINE SWITCH(A,B,IND)
      COMPLEX A(2,2),B(2,2)
      B(1,1)=A(1,1)
      B(2,2)=A(2,2)
      GO TO (2540,2570),IND
2540  B(1,2)=A(1,2)
      B(2,1)=A(2,1)
      GO TO 2590
```

2570 B(1,2)=A(2,1)
    B(2,1)=A(1,2)
2590 RETURN
    END

The call from the main program is:
    CALL SWITCH(C,A,1)
In this subroutine, A and IND come from the main program as C and 1.
The new matrix B is returned unaltered to the main program as matrix
A. When IND is transmitted as a 2, the returned matrix A is the transpose
of the matrix sent to the subroutine. As a word of explanation for this
subroutine, forming or renaming a new matrix could be avoided in the
main program, but by using this subroutine all the call statements can be
written in a redundant manner to simplify the coding.

5.14 ANALYSIS OF A PLANAR DIVIDER

The lowest order of Gysel divider (Figure 5-3), is a two-to-one divider
as shown below:

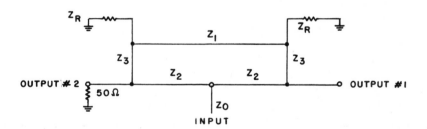

Figure 5-6

Z_2 and Z_3 are a quarter-wavelength long at midband, and Z_1, which is
two quarter-wavelengths, is one-half wavelength at midband. Z_R are
terminating loads and it is desirable for them to be equal to Z_0. The
variables are the impedance of Z_1, Z_2, Z_3, Z_R, and possibly the length
of the transformers. Redrawing the circuit (Figure 5-7a,b) to form a
cascade of components (which can be modeled by ABCD matrices) beginning at the input port, and traveling clockwise to output 1, we can see
a network that consists of this cascaded string in parallel with transformer
Z_2, the short counterclockwise path to output 1. By using subroutine
ABSDSP, the cascaded networks may be consolidated into one matrix

(IND set to 1), then the subroutine used again to put this matrix in parallel with the matrix of Z_2 to form one overall ABCD matrix between the input and output 1. From this matrix, all the divider insertion characteristics may be obtained, such as insertion loss, insertion phase, input VSWR, and output VSWR.

Along with insertion loss (from which power division and power combining efficiency are obtained), isolation between output ports is an important design parameter. By redrawing the network (Figure 5-8a,b), two parallel circuits may be obtained between output 1 and output 2.

After cascading the parallel paths into two single matrices, a parallel addition can be made to arrive at one overall ABCD matrix, which represents all the parameters of the network's isolation.

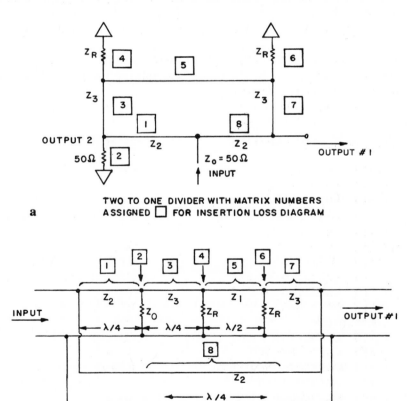

a   TWO TO ONE DIVIDER WITH MATRIX NUMBERS
     ASSIGNED ☐ FOR INSERTION LOSS DIAGRAM

b   TWO TO ONE DIVIDER INSERTION LOSS DIAGRAM (INPUT TO OUTPUT #1)
     WITH MATRIX NUMBERS ASSIGNED ☐

Figure 5-7

5.14.1 The Subroutine Sequence of the Main Program

In the main part of the computer program are the subroutine sequence calls written to obtain each particular performance desired. Using figure 5-7(b), the steps are described to obtain the insertion loss subroutine sequence.

The product of matrix 1 times matrix 2 is taken, then the product of matrix 1–2 times matrix 3 is taken. The cascading is from left to right. The final ABCD matrix is obtained by adding matrix 8 to the product matrix of 1 through 7.

Matrix 1 is a transmission line of length L and impedance Z_2. It is formed by using the subroutine ABCDXL. Matrix 2 is a shunt element representing the termination of output 2 in a normal transmission line, in this case Z_0, which is 50 ohms and is formed by using subroutine ABCDY.

ABCDSP is used to multiply matrix 1 times matrix 2. Next, subroutine SWITCH is used to change the matrix call-out, so the subroutines will always have the same call-out sequence.

The subroutine sequence to arrive at one ABCD matrix connecting the input to output 1 of the two-way divider is:

MATRIX NO.	SUBROUTINE	COMMENT
1	CALL ABCDXL(A,Z2,L,FREQ)	1/4 WAVE LINE
2	CALL ABCDY(B,Y0,2)	SHUNT ADMITTANCE
	CALL ABCDSP(A,B,C,1)	A*B=C
	CALL SWITCH(C,A,1)	C RETURNED AS A
3	CALL ABCDXL(B,Z3,L,FREQ)	1/4 WAVE LINE
	CALL ABCDSP(A,B,C,1)	A*B=C
	CALL SWITCH(C,A,1)	C RETURNED AS A
4	CALL ABCDY(B,YR,2)	SHUNT ADMITTANCE
	CALL ABCDSP(A,B,C,1)	A*B=C
	CALL SWITCH(C,A,1)	C RETURNED AS A
	DL=2.0*L	
5	CALL ABCDXL(B,Z1,DL,FREQ)	1/2 WAVE LINE
	CALL ABCDSP(A,B,C,1)	A*B=C
	CALL SWITCH(C,A,1)	C RETURNED AS A
6	CALL ABCDY(B,YR,2)	SHUNT ADMITTANCE
	CALL ABCDSP(A,B,C,1)	A*B=C
	CALL SWITCH(C,A.1)	C RETURNED AS A
7	CALL ABCDXL(B,Z3,L,FREQ)	1/4 WAVE LINE
	CALL ABCDSP(A,B,C,1)	A*B=C
	CALL SWITCH(C,A,1)	C RETURNED AS A

8   CALL ABCDXL(B,Z2,L,FREQ)   1/4 WAVE LINE
        CALL ABCDSP(A,B,C,2)   PARALLEL A AND B

Matrix C, formed in the last step (parallel), is the composite matrix from which insertion loss, insertion phase, output VSWR, and input VSWR are determined.

The program statements required are:

$$T=2.0/(C(1,1)+C(1,2)*Y0+C(2,1)*Z0+C(2,2)$$
$$TMAG=20.0*ALOG10(CABS(T))$$
$$TPHI=(ATAN2(AIMAG(T),REAL(T))*180.0/3.141596$$

T is the complex transmission coefficient, CABS is the absolute value of a complex number, and TMAG is the insertion loss in dB. ATAN2 is the arc tangent with angle quadrant designation in radians. TPHI is the insertion phase converted to degrees. Hence,

$$GAMMA=(C(1,1)+C(1,2)*YO(C(2,1)*Z0+C(2,2)))/$$
$$+(C(1,1)+C(1,2)*Y0+C(2,1)*Z0+C(2,2))$$
$$VSWR=(1.0+CABS(GAMMA))/(1.0-CABS(GAMMA))$$
$$GAMMAO=(C(2,2)+C(1,2)*Y0-(C(2,1)*Z0+C(1,1))/$$
$$+(C(1,1)+C(1,2)*Y0+C(2,1)*Z0+C(2,2))$$
$$VSWRO=(1.0+CABS(GAMMO))/(1.0-CABS(GAMMO))$$

VSWR is the VSWR at the input, and VSWRO is the VSWR at output 1, a reverse or output VSWR.

The network drawn for isolation is:

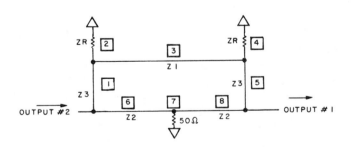

TWO TO ONE DIVIDER WITH MATRIX NUMBERS
ASSIGNED ☐ FOR ISOLATION LOSS DIAGRAM

Figure 5-8a

and the redrawn circuit for programming is:

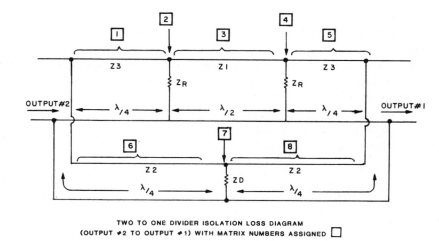

TWO TO ONE DIVIDER ISOLATION LOSS DIAGRAM
(OUTPUT #2 TO OUTPUT #1) WITH MATRIX NUMBERS ASSIGNED ☐

Figure 5-8b

As in the insertion loss case, the matrix products are taken for the series strings, then the two parallel matrices are added.

MATRIX NO.	SUBROUTINE	COMMENTS
1	CALL ABCDXL(A1,Z3,L,FREQ)	1/4 WAVE LINE
2	CALL ABCDY(B1,YR,Z)	SHUNT ADMITTANCE
	CALL ABCDSP(A1,B1,C1,1)	A1*B1=C1
	CALL SWITCH(C1,A1,1)	C1 RETURNED ASA1
3	CALL ABCDXL(B1,Z1,DL,FREQ)	1/2 WAVE LINE
	CALL ABCDSP(A1,B1,C1,1)	A1*B1=C1
	CALL SWITCH(C1,A1,1)	C1 RETURNED AS A1
4	CALL ABCDY(B1,YR,2)	SHUNT ADMITTANCE
	CALL ABCDSP(Z1,B1,C1,1)	A1*B1=C1
	CALL SWITCH(C1,A1,1)	C1 RETURNED AS A1
5	CALL ABCDXL(B1,Z3,L,FREQ)	1/4 WAVE LINE
	CALL ABCDSP(A1,B1,C1,1)	A1*B1=C1 (HOLD C1)
6	CALL ABCDXL(A1,Z2,L,FREQ)	1/4 WAVE LINE
7	CALL ABCDY(B1,Y0,2)	SHUNT ADMITTANCE
	CALL ABCDSP(A1,B1,C2,1)	A1*B1=C2
	CALL SWITCH(C2,A1,1)	C2 RETURNED AS A1
8	CALL ABCDXL(B1,Z2,L,FREQ)	1/4 WAVE LINE
	CALL ABCDSP(A1,B1,C2,1)	A1*B1=C2
	CALL ABCDSP(C1,C2,C3,2)	PARALLEL C1 AND C2

C3 is the resulting matrix and the isolation is found by evaluating the complex transmission coefficient T as in the insertion loss case. The isolation is the common log of the absolute magnitude of T times 20.

In order to implement a program, it is necessary to define complex numbers, arrays, and input design values, as well set up the frequency variable parameters. In this program, the complex values are defined as:

COMPLEX A(2,2), B(2,2),C(2,2),T,GAMMA
COMPLEX A1(2,2, B1(2,2),C1(2,2),C2(2,2),C3(2,2)
COMPLEX T1, GAMMAO

The required input data is:
Z_0, Z_1, Z_2, Z_3, and Z_R, in ohms; midband frequency in megahertz; low frequency, high frequency, in megahertz; and the number of frequency steps.

As formatted, this program will only predict frequency behavior for a given set of fixed line impedances, line lengths, and terminating loads. With a few modifications, the program could vary, within selected ranges, some of the fixed parameters to explore network sensitivity.

5.14.2 Analysis of a Five-to-One Planar Divider

A five-to-one planar Gysel divider may be analyzed using the same network programming method as in the two-to-one divider. The five-to-one divider is shown schematically as:

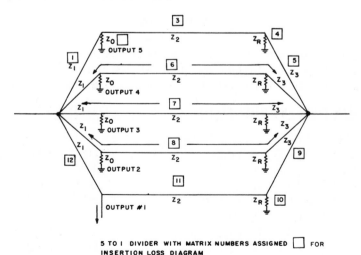

5 TO I DIVIDER WITH MATRIX NUMBERS ASSIGNED ☐ FOR INSERTION LOSS DIAGRAM

Figure 5-9

where Z_1, Z_2, and Z_3 are one-quarter wavelength long at midband. Z_R are terminating loads and are ideally equal to Z_0. To calculate the insertion loss from the input port to output 1, the network is redrawn as shown with Z_0 terminations at outputs 2 through 5.

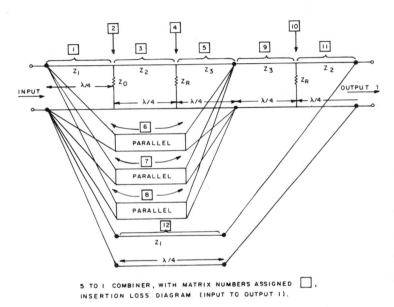

5 TO 1 COMBINER, WITH MATRIX NUMBERS ASSIGNED ☐,
INSERTION LOSS DIAGRAM (INPUT TO OUTPUT 1).

Figure 5-10

Again, the main body of the program consists of the subroutine sequence that is used to find the final ABCD matrix, from which the insertion loss and other parameters are determined. This is listed below:

MATRIX NO.	SUBROUTINE	COMMENTS
1	CALL ABCDXL(A,Z1,L,FREQ)	1/4 WAVE LINE
2	CALL ABCDY(B,Y0,2)	SHUNT ADMITTANCE
	CALL ABCDSP(A,B,C,1)	A*B=C
	CALL SWITCH(C,A,1)	C RETURNED AS A
3	CALL ABCDY(B,Z2,L,FREQ)	1/4 WAVE LINE
	CALL ABCDSP(A,B,C,1)	A*B=C
	CALL SWITCH(C,A,1)	C RETURNED AS A
4	CALL ABCDY(B,YR,2)	SHUNT ADMITTANCE
	CALL ABCDSP(A,B,C,1)	A*B=C
	CALL SWITCH(C,A,1)	C RETURNED AS A

5	CALL ABCDXL(B,Z3,L,FREQ)	14 WAVE LINE
	CALL ABCDSP(A,B,C,1)	A*B=C
	CALL SWITCH(C,A,1)	C RETURNED AS A
	CALL SWITCH(A,B,1)	A RETURNED AS B
6	DO 1130 KK=1,K	(K=3 IN THIS CASE)
7	CALL ABCDSP(A,B,C,2)	A+B=C
8	CALL SWITCH (C,A,1)	C RETURNED AS A
1130	CONTINUED	(MATRIX B HAS BEEN REPEATEDLY ADDED TO MATRIX A)
9	CALL ABCDXL(B,Z3,L,FREQ)	1/4 WAVE LINE
	CALL ABCDSP(A,B,C,1)	A*B=C
	CALL SWITCH(C,A,1)	C RETURNED AS A
10	CALL ABCDY(B,YR,2)	SHUNT ADMITTANCE
	CALL ABCDSP(A,B,C,1)	A*B=C
	CALL SWITCH(C,A,1)	C RETURNED AS A
11	CALL ABCDXL(B,Z2,L,FREQ)	1/4 WAVE LINE
	CALL ABCDSP(A,B,C,1)	A*B=C
	CALL SWITCH(C,A,1)	C RETURNED AS A
12	CALL ABCDXL(B,Z1,L,FREQ)	1/4 WAVE LINE
	CALL ABCDSP(A,B,C,2)	A+B=C

C is the resulting matrix from which we obtain insertion loss and all the other parameters. Notice that the DO loop at matrix level 6,7,8 sets the number of branches, so that by adjusting K the number of branches can be modified without rewriting the program.

The isolation between outputs may be calculated by again using a redrawn network.

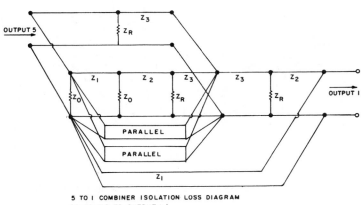

5 TO I COMBINER ISOLATION LOSS DIAGRAM
(OUTPUT 5 TO OUTPUT I)

Figure 5-11

Here we have a problem. Note that the three parallel circuits connect across the middle of the network, prevent the usual combining in series, parallel, or cascade. However, the circuit is three quarter-wavelengths long, so a high degree of decoupling occurs. For the sake of simplicity, with this path removed, the two-port circuit analysis may be applied.

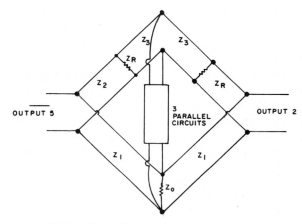

5 TO I COMBINER ISOLATION LOSS DIAGRAM
(OUTPUT 5 TO OUTPUT I) REDRAWN

Figure 5-12

The appropriate circuit is:

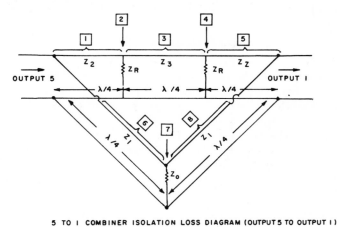

5 TO I COMBINER ISOLATION LOSS DIAGRAM (OUTPUT 5 TO OUTPUT I)
SIMPLIFIED WITH MATRIX NUMBERS ASSIGNED ☐

Figure 5-13

The subroutine calls to define isolation using this network are:

MATRIX NO.:	SUBROUTINE	COMMENTS
1	CALL ABCDXL(A,Z2,L,FREQ)	1/4 WAVE LINE
2	CALL ABCDY(B,YR,2,)	SHUNT ADMITTANCE
	CALL ABCDSP(A,B,C,1)	$A*B=C$
	CALL SWITCH(C,A,1)	C RETURNED AS A
		$DL=2.0*L$
3	CALL ABCDXL(B,Z3,DL,FREQ)	1/2 WAVE LINE
	CALL ABCDSP(A,B,C,1)	$A*B=C$
	CALL SWITCH(C,A,1)	C RETURNED AS A
4	CALL ABCDY(B,YR,2)	SHUNT ADMITTANCE
	CALL ABCDSP(A,B,C,1)	$A*B=C$
	CALL SWITCH(C,A,1,)	C RETURNED AS A
5	CALL ABCDXL(B,Z2,L,FREQ)	1/4 WAVE LINE
	CALL ABCDSP(A,B,C1,1)	$A*B=C1$
		PRODUCT HOLD C1
6	CALL ABCDXL(A,Z1,L,FREQ)	1/4 WAVE LINE
7	CALL ABCDY(B,YO,2)	SHUNT ADMITTANCE
	CALL ABCDSP(A,B,C,1)	$A*B=C$
	CALL SWITCH(C,A,1)	C RETURNED AS A
8	CALL ABCDXL(B,Z1,L,FREQ)	1/4 WAVE LINE
	CALL ABCDSP(A,B,C,1)	$A*B=C$
	CALL ABCDSP(C1,C,A,2)	$C1+C=A$

A is the resulting matrix, from which the insertion loss, in this case isolation loss, may be calculated.

REFERENCES

1. Foti, S. J., R. P. Flam, and W. J. Scharpf, "60-Way Radial Combiner Uses No Isolators," *Microwaves/RF*, Vol. 23, No.7, July 1984, pp.96–118.
2. Sanders, B. J., "Radial Combiner Runs Circules Around Hybrids," *Microwaves,* Nov. 1980, pp.55–58.
3. Howe, H., *Stripline Circuit Design,* Artech House, Dedham, MA, 1974.
4. Wilkinson, E. J., "An N-Way Hybrid Power Divider," *IEEE Trans. Microwave Theory Tech.,* Vol. MTT-8, No.1, Jan.1969, pp.116–118.
5. Parad, I. I., and R. L. Moynihan, "Split Tee Power Divider," *IEEE Trans. Microwave Theory Tech.,* Vol. MTT-13, No. 1, Jan. 1965, pp.91–95.

6. Cohen, S., "A Class of Broadband, Three-Port to TEM Mode Hybrids," *IEEE Trans. Microwave Theory Tech.*, Vol. MTT-16, No.2, Feb. 1968, pp.110–118.

7. Gysel, U., "A New N-Way Power Divider/Combiner Suitable for High Power Applications," 1975 IEEE MTT-S International Microwave Symposium Digest, pp.116–118.

8. White, J. F., *Semiconductor Control*, Artech House, Dedham, MA, 1977. Appendix E, pp.521–525.

9. *Ibid.*, Chapter 6, pp.177–243.

10. Richards, P. I. , "Applications of Matrix Algebra to Filter Theory," *Proc. IRE Wave Electrons*, March 1946, pp.145–150.

11. Gupta, K. C., *et al., Computer Aided Design of Microwave Circuits*, Artech House, Dedham, MA, 1981.

Chapter 6
Power Amplifier Modules

Solid-state module designs generally fall into one of two common categories: (1) transmit-receive modules for radiating multi-element phased array radars, and (2) transmit-only modules for single-antenna "bottle" transmitters [1]. Functional block diagrams for each of these module designs are shown in Figure 6-1. The transceiver module is a compact subsystem containing a receiver front end, transmitting amplifier T/R switching, low-noise amplifier and limiter, and BITE and logic circuits. The transceiver module feeds a radiating element and receives its own RF device from a corporate feed, or a space-fed lens. The transmit-only module contains just the amplifier and *built-in test equipment* (BITE) of the transceiver module, but can be equally complex. Modern system parameters that call for wide bandwidths, large average power requirements, high packaging densities, and fault-tolerant operation with 100 percent availability make for challenging electromechanical designs.

The baseline plan for the solid-state transmitting amplifier is to provide as large a segment of transmitted power as practical for the application. Because the capabilities of power transistors range from 100 to 1000W for UHF through S band, practical and reproducible modules with power capabilities from 0.5 to 5kW are common [2–8]. High power output is achieved by placing identical devices in parallel using microwave combining techniques to maintain low load VSWR, and provide for stage-to-stage isolation [9]. Additionally, a successful design allows for practical manufacturing with realistic variations from one device to another, adequate cooling mechanisms, suitable RF drive distribution, built-in test and protection, and allowances for easy troubleshooting and repair.

A fundamental rule in class C amplifier design is to maintain low VSWR in the module. Measured data for the characterization of a device at 1350 MHz will serve as reference for an imaginary module and single-stage design. The data are given in Figure 6-2. Note that maximum output power is achieved at a load impedance of 5.0 $-j3.5$ ohms. Assuming a mature device and that the T_j distribution has been minimized, a desired load impedance of 6.0 $-j3.0$ ohms is chosen. From the characterization data, this should yield the following performance parameters at 1350 MHz:

- Power output = 105W
- Power input = 20W
- Collector efficiency = 52%
- Peak T_j = 103 degrees at 30 degrees sink
- Gain = 7.1 dB

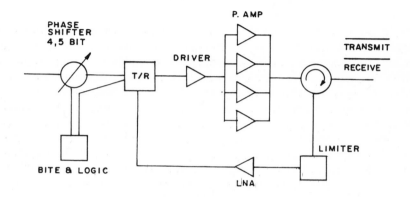

BLOCK DIAGRAM : TRANSCEIVER MODE

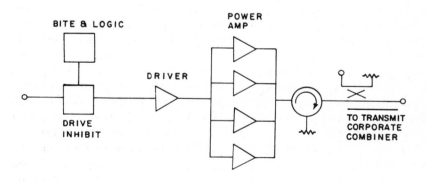

BLOCK DIAGRAM: TRANSMIT AMPLIFIER MODULE

Figure 6-1

The following potential scenario may now occur: the circuit is designed and provides the anticipated target load impedances; two dozen single-stage circuits are built and tested; the manufacturer verifies and commits to the performance; the single-stage amplifier is configured in parallel in a module, combined, and total power output from the module is measured 1 dB below the anticipated output power level.

The output parameters of class C devices exhibit extreme sensitivity to load VSWR. Consequently, a well-defined load impedance must be designed [10]. Typical insertion loss numbers are shown in Figure 6-3. Assuming that device performance is not dependent on load VSWR, the expected performance would be that shown. Individual power-combining networks and auxiliary module elements (circulator, connector and cables, monitor couplers, *et cetera*) may have low individual reflection coefficients, but when cascaded in series and additive in phase, load VSWRs on the order of those shown in Figure 6-4 may result at individual output stages. The net effect of the increased VSWR, which each final amplifier stage exhibits, is a degradation of power output below the design level of 105W. Additionally, the input power divider has finite return loss and pulls the load on the driver device. The load pulling effect of the input power divider can degrade the power output of the driver device so that the RF drive level to each of the final stages is less than the nominal level.

Extreme care must be taken in order to ensure that the load impedance of the cascaded microwave circuitry to the output of the final stages be kept as low as possible. Circulators that are used on the output of the module to maintain a nearly constant load VSWR may exhibit load impedances as high as 1.25:1 over 15 percent bandwidth and as high as 1.5:1 over a 60 percent bandwidth. A small impedance-matching network, empirically placed between the circulator and the module power combiner, may help maintain performance levels. Also, the active match of the power combiner used to pool the power outputs of the parallel stages can be measurably different from the two-port small-signal measurements on the network analyzer. Figure 6-5 represents measured passive matches on a wideband 4:1 stripline combiner in comparison with the active match.

Insertion phase dependence is another source of output power degradation due to load mismatch [11]. As the load contour information of Figure 6-2c depicts, the relative insertion phase of a device is at the mercy of variations in the reflection coefficient. The relative insertion phase measured for this device varied from a total δ of 40 degrees over a VSWR range of 1.45:1. When power levels of this phase δ are combined, an overall dB loss is passed on to dummy loads:

$$\text{dB lost} = 20\log\left[\frac{1+\cos\theta_e}{2}\right] \qquad (6\text{-}1)$$

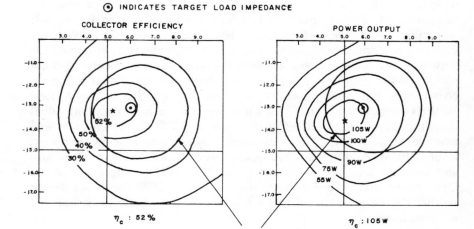

COLLECTOR LOAD CONTOURS

⊙ INDICATES TARGET LOAD IMPEDANCE

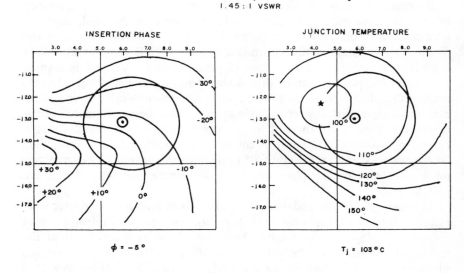

Figure 6-2

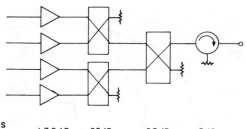

GAIN/LOSS
BUDGET :   +7.2 dB   --.25dB   -.25 dB   -.5 dB

POWER
LEVEL :   (x4)20   (x4)105   (x2)198W   374W   334W

TYPICAL LOSSES AT 1350 MHz

Figure 6-3

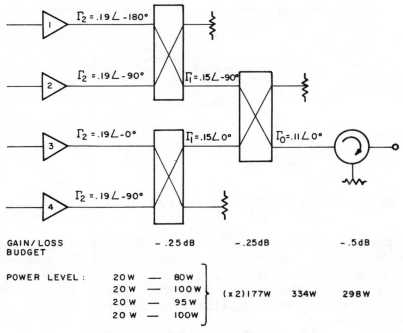

GAIN/LOSS   -.25 dB   -.25dB   -.5dB
BUDGET

POWER LEVEL :   20 W ── 80W ⎫
                 20 W ── 100W ⎪
                 20 W ── 95 W ⎬ (x2)177W   334W   298W
                 20 W ── 100W ⎭

LOAD PULLING EFFECT OF MODULE OUTPUT.
VSWR ON POWER OUTPUT OF FINAL STAGES

Figure 6-4

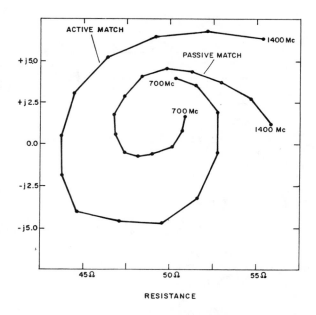

RESISTANCE

MEASURED IMPEDANCE DIFFERENCES BETWEEN PASSIVE
ONE PORT EXCITATION & ACTIVE MULTIPLE EXCITATION ON
A QUAD HYBRID

Figure 6-5

where

θ_e = phase combining error

$$\text{dB lost} = 10\log\left[\frac{1+\cos(40°)}{2}\right] = .54 \text{ dB}$$

In this case .54 dB power output conceivably could be lost to phase combining errors. This situation is not uncommon. Multiple stages of 3-dB quadrature hybrids are often used as mirror image splitting and combining networks. When used in conjunction with the insertion phase characteristics of class C amplifiers, significant phase combining errors could accrue.

A property of the quadrature hybrid (Figure 6-6) is that it provides an input impedance match that is degraded only by its isolation when the two output ports are terminated in equal mismatches. Reflected power is combined in phase in the dummy load at port 2. The net effect is to maintain a relatively constant, high quality impedance match to the output of the driver transistor. However, the mirror image of the divider must be used on the output. Load mismatches on the output will be reflected back toward the output transistors such that the output transistors see the same magnitude reflection coefficient scattered at 90 degree intervals.

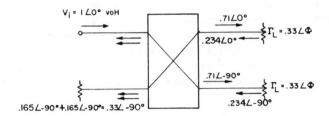

PHASED VOLTAGES IN A QUADRATURE HYBRID

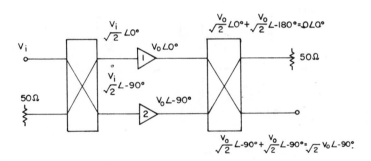

HYBRID COUPLED AMPLIFIER AMPLIFIER PAIR

Figure 6-6

The cumulative pushing and pulling effects on power output and insertion phase are shown in Figure 6-7. Additionally, the insertion phase of a transistor can change due to phase pushing effects, i.e., class C devices demonstrate insertion phase changes in the relative RF drive level. Measured L band devices have shown 10–15 degrees/dB phase sensitivity to drive and 0.2–0.5 dB/dB amplitude sensitivity to drive [12]. Amplitude sensitivity depends on how far into saturation the device is operated. In a hybrid coupled amplifier, the drive levels to each of the final stages may vary from the nominal due to the coupling characteristics of the divider. This situation is only aggravated for wide bandwidth applications as shown in Figure 6-8. For ultra-wideband applications (>60%), coupled structures described by Cohn [13] alleviate some of the problems of phase and amplitude pushing, but at the expense of insertion loss. For a relatively narrowband application of 15 percent, amplitude variations can be held to ± .10 dB per tier of hybrid combining, and, for 50 percent bandwidth, amplitude variations of ± .35 dB can be expected for single quarterwave structures [14]. It is obviously wise to use in-phase split-tee Wilkinson combiners for most of the combining, and use one level of quadrature combining to provide the isolation between the driver and the final stages. Schemes that use an offset 90 degree line in conjunction with in-phase power dividers are an inexpensive and practical implementation of the quadrature concept over very narrow bandwidths.

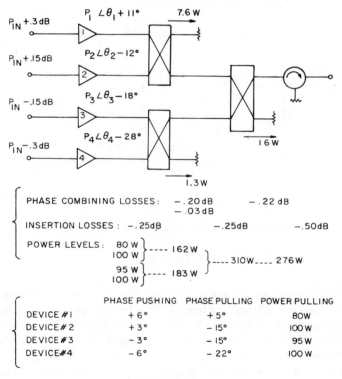

Figure 6-7

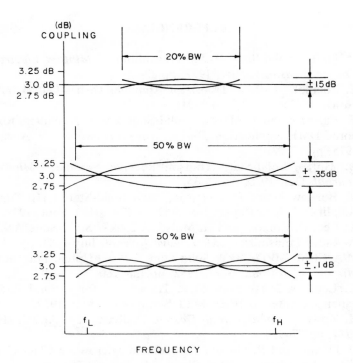

QUADRATURE HYBRID COUPLING CHARACTERISTICS

Figure 6-8

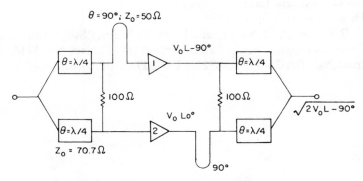

HYBRID COUPLED PAIR USING
90° OFFSET ARMS & WILKINSON COMBINERS

Figure 6-9

REFERENCES

1. D. Hoft, "Solid-State Radar Transmitters," *Military Electronics/ Countermeasures,* Nov. 1977, pp.22–26.
2. E. Ebersol, "L-Band Transistor Amplifier Dishes Out 1kW," *Microwaves,* Dec. 1972, pp.9–21.
3. H. Balshem and T. Marks, "Solid-State Power Amplifiers for Airborne DME Applications," *Microwave Systems News,* April/May 1975, pp.17–23.
4. D. Hoft, "Solid-State Transmit/Receive Module," *Microwave Journal,* Oct. 1978, pp.33–35.
5. M. Borkowski and D. Laighton, "2kW Solid-State UHF Transmit Amplifier," Final Report No. N62269-77-C-0420, Sept. 1978.
6. K. Lee, C. Corson, and G. Mols, "A 250kW Solid-State AN/SPS-40 Radar Transmitter," *Microwave Journal,* July 1983, pp.93–105.
7. M/A-COM, MPD, "1kW Solid-State L-Band Uplink Transmitter," *Microwave Journal,* Sept. 1982, pp.158–160.
8. K. Lee, "A 25kW Solid-State Transmitter for L-Band Radars," paper presented at IEEE MTT Symposium, May 1979.
9. H. Howe, *Stripline Circuit Design,* Dedham, MA, Artech House, 1974, pp.77–110.
10. O. Pitzalis and R. Gilson, "Broadband Microwave Class-C Transistor Amplifiers," *IEEE Trans. Microwave Theory Tech.,* vol. MTT-21, no.2, Feb. 1970, pp.108–119.
11. L. R. Lavallee, "Two-Phased Transistors Shortchange Class-C Amps," *Microwaves,* Feb. 1975, pp. 48–54.
12. Borkowski and Laighton, *op. cit.*
13. Howe, *op. cit.,* pp.111–180.
14. S. B. Cohn, P. M. Sherk, and E. M. T. Jones, "Strip Transmission Lines and Components," Final Report SRI Project No. 1114, Contract No. DA-36-039-sc-63232, Feb. 1957, pp.64–78.

Chapter 7

Power Supply and Energy Storage Design

The large signal class C amplifiers in a *solid-state transmitter* are sensitive to power supply voltage variation. The output signal from each unit amplifier will change in amplitude and phase as the power supply voltage varies. This can have an effect on the losses in the output power combining network, if one is used, or cause equivalent losses in a phased array due to element signal phase and amplitude errors. Power supply voltage fluctuation can also produce pulse-to-pulse amplitude and phase variation, which limit the performance of *moving target indicator* (MTI) circuits. Finally, the choice of the power supply and its method of power distribution can have great influence on the reliability and availability of the transmitter. These factors are also related to acquisition cost and the cost of maintenance and logistical support. For example, a system in which a single, large power supply with regulator is employed has a lower acquisition cost than a system that employs eight power supplies to give the same output. A single power supply failure disables the transmitter in the first case, while the second design has a power output decrease of slightly more than one dB and can continue operation until a spare power supply can be installed during some convenient maintenance operation.

7.1 POWER SUPPLY OPTIONS

The choice of a power supply and regulator is driven by both electrical and environmental requirements. Pulsed transmitters, which are typical in most radar systems, require the use of an energy storage capacitor to deliver the very high peak current during the pulse. The power supply and associated regulator circuits are active during the interval between pulses when the energy storage capacitor is recharged in order to be ready for the next pulse. The regulator circuit recharges the capacitor at a nearly constant current until the voltage reaches a preset level and the charging current is reduced to zero. It is desirable to set the charging current to a value that just reaches the desired capacitor voltage at the time the next pulse is to start. Otherwise, in high power transmitters the transient input current changes can cause a large pulsating load on the primary power source, usually called "thump." The regulator can be active during the pulse interval, but its effect will be small because the charging current is small compared to the pulse current.

Regulator circuits that employ a power transistor in series with the load (series-pass type) are widely used. Integrated circuit regulators are available, packaged with a series-pass power transistor. The control circuits and power transistors are also packaged separately in many applications, particularly when the power supply voltage required is not one of the standard computer circuit voltages. High performance integrated circuits are also available to provide proper control pulses and timing for switching regulators. The switching regulator is more efficient than the series-pass regulator. It also has lower volume and weight for a given power output rating than the series-pass regulator. It is the best choice for airborne and spacecraft requirements where size, weight, and efficiency are important. The switching regulator is more difficult to design and troubleshooting is a challenge in these systems. The cost of the switching regulator is usually higher than that of the less efficient analog series-pass design. Pulse transients may present more of a problem with the switcher. Reliability should be better with high efficiency because this implies lower junction temperatures, but recent experience with switchers shows this not to be so. When a switcher fails, the fault causes a chain of failures in components that otherwise would-have operated reliably. The efficiency and weight advantages of the switcher should lead to continued improvement in performance and reliability in the future.

7.1.1 Effect of Power Supply Size on Reliability

The choice of the RF power-amplifier module size has an effect on the requirement for the power supply. We could provide a single power supply to provide for the entire transmitter, or, at the other extreme, provide a single power supply for each RF module. If the components are stressed at the same level in each case, the single power supply should have a much lower failure rate than the ensemble of power supplies. On the other hand, a single failure in the ensemble of power supplies will have a negligible effect on the overall transmitter output level. The choice is further complicated by cost considerations. There may be an optimum design for which the reliability is high and cost is acceptable. The effect of a single power supply unit failure on the transmitter output is shown in Figure 7-1.

The choice of a power supply/regulator configuration is further complicated by environmental considerations that include the cooling system medium and the expected variation in ambient temperature, the quality of the raw power source, and exposure to other hostile environmental elements such as salt spray, solar flux, and unskilled maintenance personnel.

The basic structure of the transmitter is one that starts with a single signal source and ends with a large number of power amplifier modules that have their output added coherently. The power supply and distribution network must avoid failure modes in which drive power to the early stages of the corporate structure can disable a large number of output stages. A small number of redundant backup power supplies and driver modules for these critical stages is a good investment.

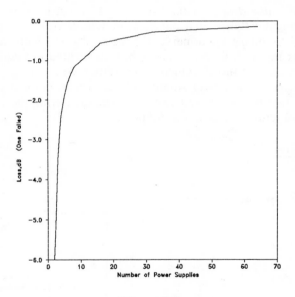

Figure 7-1

7.2 PULSE WAVEFORM AMPLITUDE AND PHASE VARIATION

The variation of output pulse amplitude with power supply voltage is very predictable. The amplitude tracks the supply voltage linearly over a fairly wide range, at least as wide as the supply voltage fluctuation. A 10 percent decrease in power supply voltage reduces the output pulse amplitude to 90 percent of the original amplitude and the change is 20 log (.9) = −0.92 dB. The voltage sensitivity of output amplitude is often expressed in dB per volt. With a nominal supply voltage of 35 volts, the sensitivity to a one volt change is 20 log (34/35) or −0.25 dB/volt.

The phase sensitivity is not as predictable and is based on empirical measurements of phase change as the power supply voltage is varied. The phase sensitivity of the PAVE PAWS module is typically 2.5 degrees per volt. This parameter is dependent on the operating frequency, the number of active elements in the module, and probably the type of transistors used.

7.3 DROOP IN PULSED TRANSMITTERS

We define *pulse droop* as the normalized variation of the pulse top from a rectangular flat-topped pulse. This definition is not yet included in the IEEE Standard Dictionary Of Electrical and Electronics Terms. The intent of the term as used here is best illustrated in Figure 7-2. It is important because of the aforementioned amplitude and phase sensitivity of the RF pulse to the power supply voltage, which in most cases is the voltage supplied by the energy storage capacitor during the pulse. With reference to Figure 7-2, droop is defined as

$$d = \frac{E_1 - E_2}{E_1} \tag{7-1}$$

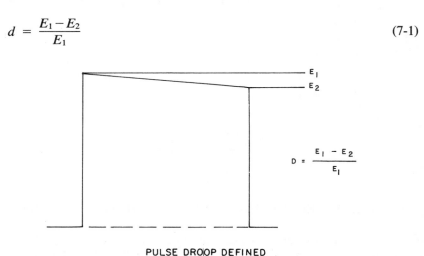

PULSE DROOP DEFINED

Figure 7-2

It can be expressed in percent or dB, as desired. The energy storage capacitor and the RF module pulse-load can be modelled as a capacitor in parallel with a resistor and switch connected in series. The switch is closed during the pulse. Usually, the droop is small enough to allow a linear approximation to the voltage *versus* time portion of the exponential to be valid. With a supply voltage V_{cc} at the start of the pulse, the voltage droop is

$$d = \frac{I_p T_p}{C V_{cc}} \tag{7-2}$$

where I_p is the net discharge current during the pulse and T_p is the pulse duration. In low duty cycle systems, the regulator charging current is negligible compared to the discharge current delivered to the RF module.

This equation can be used to find the size of the energy storage capacitor required in order to keep droop below an acceptable level. The capacitor size is

$$C = \frac{I_p T_p}{d V_{cc}} \tag{7-3}$$

For example, with a peak pulse current of 100 amperes and a supply voltage of 35 V, the capacitor required to hold droop to 2 percent (.02) with a 100 microsecond pulse is

$$C = (100 \times 100)10^{-6} \times (0.02)35$$

or

$$C = 14,286 \text{ microfarads } (\mu F)$$

The energy stored in the capacitor is

$$E_s = 1/2 \ C \ V_{cc}^2 = \frac{I_p V_{cc} T_p}{2d} \tag{7-4}$$

The output pulse energy is

$$E_o = \eta \frac{I_p \ V_{cc} \ T_p}{100} \tag{7-5}$$

η is the module collector efficiency in percent. The ratio of the energy stored in the capacitor to the RF output pulse energy is

$$E_s/E_o = \frac{100}{\eta\, 2d} \tag{7-6}$$

For example, with a collector efficiency of 50 percent and a droop of 2 percent, the energy stored would have to be 50 times the energy in the output pulse.

A more complete picture of the collector voltage waveform would include the *equivalent series resistance* (ESR) of the energy storage capacitor and its lead inductance. The high pulse currents are best met by computer-grade electrolytic capacitors. The best quality units are rated for high-temperature service and are constructed for low ESR. It is important to keep the connecting wires short and use a wire that will provide low ohmic drop during the pulse. To illustrate this point, consider an RF module that provides a peak output power of 400W at an efficiency of 40 percent. The collector supply is nominally 32V and a 20,000μF capacitor is connected to the module via a pair of one foot long No. 10 AWG annealed copper-wire leads. The capacitor has an ESR rated at 8 milli-ohms. The two feet of lead wire and the internal module stripline resistance adds an additional 4 milli-ohms. The module peak current is

$$I_p = (400/32)0.4 = 31.25 \text{ amperes}$$

The voltage drop in the ESR and connections to the capacitor is $31.25 \times 12 \times 10^{-3} = 0.375\text{V}$. This is typical of a careful design. The drop can be considerably worse if the capacitor is located at a mechanically good but electrically poor distance from the module, and a high resistance path is provided.

7.4 MTI IMPROVEMENT FACTOR LIMITATIONS

The amplitude and phase instability in a train of output RF pulses place an upper limit on the MTI *improvement factor* that can be realized in a radar system. The limiting improvement factor due to phase instability is

$$I = 20\log\left(\frac{1}{\Delta\phi}\right) \tag{7-7}$$

The corresponding upper limit on improvement factor caused by amplitude instability is

$$I = 20\log\left(\frac{A}{\Delta A}\right) \qquad\qquad (7\text{-}8)$$

When a power supply regulator is employed as a charge source for the energy storage capacitor, it provides a constant current during the interpulse interval, which should just charge the capacitor to the design level at the start of the next pulse. If the interval between pulses is not uniform, the voltage at the start of each pulse will vary in proportion to the interpulse interval. This can lead to both amplitude and phase variation, although the amount of this variation should be accurately predictable. This predictable variation could be compensated for in the signal processing circuits. Unpredictable amplitude and phase variation will limit the MTI performance in accordance with the above equations. This type of variability can arise when the exciter frequency source is unstable or when the power supply regulator is not stable. A uniform interpulse interval for the waveforms used in MTI radars should simplify signal processing. If a burst of uniformly spaced, equal duration pulses is employed for MTI, then the amplitude and phase at the start of each pulse will either be constant or have a constant rate of change. The latter case prevails when the interpulse interval is smaller than the average over the complete waveform cycle; for example, when a burst of MTI pulses of uniform duration and spacing is combined with other waveforms. In either case, a three-pulse canceller circuit will cancel the linear amplitude and phase changes from pulse to pulse over a wider range than we might predict from a literal interpretation of (7-7) and (7-8). This can bring about savings in size and cost of the energy storage capacitors and power supply regulators. Similar savings are possible when linear FM pulse compression is employed. If the phase of the RF output runs down with decreasing collector voltage, a *down chirp* transmitting waveform has additional phase change added as the energy storage capacitor is discharged while an *up chirp* waveform must overcome the down slope caused by the capacitor discharge. For a given required phase slope, a down chirp waveform can be implemented with a smaller energy storage capacitor bank and a smaller power supply regulator.

Chapter 8

Reliability of Solid-State Transmitters

Reliability is defined as the probability that a system performs its intended function over a given time period. Reliability can be related to the *mean time between failures* (MTBF) by

$$R = e^{-t/MTBF} \qquad (8\text{-}1)$$

This equation is valid during the time period when the system has been "burned in" and failures are random in nature. Each component in the system has a predicted MTBF, which depends on the type of component and its operating environment. The component failure rate is defined as the reciprocal of its MTBF. In a system in which the reliability is equal to the product of its combined component reliabilities, the exponential expression that relates reliability to the failure rate for each component leads to the result that the system failure rate is equal to the sum of the component failure rates. An excellent source of component failure rate data is MIL-HDBK-217D, which is periodically updated by Rome Air Development Center (RBE-2), Griffiss Air Force Base, New York. It provides (in section 5.1.3.9) failure rate data on bipolar microwave transistors as a base failure rate of 0.1 failures/10^6 hours multiplied by a number of factors, including the quality factor, application factor, peak power and frequency, temperature, matching network, and environment factors. When all of these factors are applied, the base failure rate can easily increase by several orders of magnitude. The lowest failure rate factors are obtained with JANTXV quality devices, which have been IR scanned for die attachment and screened for barrier layer pinholes on gold metallized devices, operated as a pulsed amplifier at less than 5 percent duty cycle, with a peak operating power of less than 100W at a frequency in the 0.2 to 0.4 GHz range and a junction temperature of 100°C, a collector voltage to breakdown voltage ratio of 40 percent, with both input and output matching, and a ground benign or space orbital environment. This combination would have a base failure rate derating factor of 0.1, and a transistor failure rate of 0.01/10^6 hours.

MIL-HDBK-217D also gives failure rates for *aluminum electrolytic* capacitors as a product of voltage derating, quality, capacitance, and environmental factors. The energy storage capacitor and RF transistor are the most critical components that affect system reliability. The failure rate data presented may contain errors because of the time lag in gathering the data as technological improvements take place during the same period. The absolute value of the failure rate is not as important as the relative values obtained as derating factors are varied. A ten-degree increase in junction temperature, for example, gives a 40 percent increase in failure rate for *refractory metal-gold* RF transistors. Awareness of the effect on reliability of changes in component derating make the process worthwhile, especially during the early stages of transmitter design when changes are more affordable.

There are some special properties that make the solid-state transmitters more reliable than an equivalent single vacuum-tube powered transmitter. The most important feature is the very low probability that the occurrence of a random failure will cause the transmitter to cease operating in a satisfactory manner. A second and equally important property is the slow rate at which a failure state is approached. Total and sudden failure is a very rare mode of failure, and the more likely failure modes cause a gradual and predictable reduction of power output over time of operation. These qualities, all too often described as "fault tolerance" and "graceful degradation," are the reason why a solid-state transmitter with an initial acquisition cost that is higher than that of a single tube-powered transmitter is potentially a better investment. Maintenance can be deferred to a time when the cost and penalties of the maintenance action are minimized. Logistical costs are reduced because a small RF transmit module is light in weight, low in cost, and easily handled by a single technician, while a high-power *traveling wave tube* or *klystron* represents a large investment that occupies a lot of space and must be replaced by a whole crew of technicians.

The following material is intended to supplement the reliability engineering data that is available in many publications [1,2,3]. The reliability prediction of the CSA transmitter is described. Some related cost trade-off data are also given to illustrate the value of fault tolerant systems over those that go dead all at once.

8.1 RELIABILITY PREDICTION FOR A CSA

The first stage of a CSA is in series with the rest of the amplifier, and a failure at this stage will render the transmitter useless. While the probability of a failure at this point is very low, the consequences of a failure are very high and we are well motivated to add redundant backup at this stage in order to avoid a catastrophic failure, which will rarely occur. At the next rank of the CSA, the fan-out to the multiple rank one stages reduces the effect of the failure of any one element although the probability of failure for a member of this rank has increased in proportion to the increase in its population. A reliability block diagram can be constructed as a serial string of the CSA ranks, each of which has a population G times the one that drives it. If the CSA is built up of unit amplifiers with gain G and failure rate λ, then the failure rate of each higher rank block increases by G over the block to its left as shown in Figure 8-1. The corresponding MTBF is shown on each block. In a UCSA, or a CSA in which each unit amplifier is identically stressed, the probability of any unit amplifier failing is the same regardless of the location of the amplifier in the transmitter. The probability that a failure occurs within a given rank is proportional to the ratio of the population of that rank to the total transmitter population. A failure at the input stage (rank 0) causes all drive to the transmitter to be lost. At the next stage, the probability of failure has increased by G, but the effect of a single failure is reduced to $(G-1/G)^2$. This process continues as the population of each rank grows rapidly and the effect of a single failure becomes negligible. What is needed is a specified level of power output degradation below which the transmitter is considered to have failed. It is then possible to calculate the amount of operating hours that can be expected before that level of degradation occurs. Periodic maintenance visits to replace failed modules can be scheduled so as to minimize the probability that the transmitter power reaches the specified threshold during the interval between maintenance visits, which at the same time minimizes the cost of maintenance.

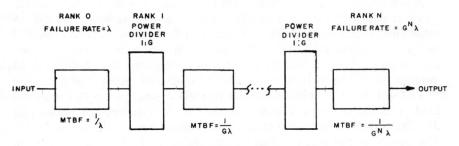

C.S.A. RELIABILITY BLOCK DIAGRAM

Figure 8-1

The design of the transmitter will be affected by the failure modes and affects analysis. Redundant elements will be introduced at rank levels, which require this protection regardless of the low failure probability. The effect of a single failure depends on the rank in which the failure occurs. The number of devices in a rank depends on the power output rating of the transmitter and the power rating of the devices employed. We can compute tables which list the number of devices that can fail before the transmitter has its performance degraded below an acceptable level. For the output rank, the percentage of failures is related to the power output reduction in dB as follows:

$$\text{Power loss} = 10\log\ [N_s/N_o]^2 \text{dB} \qquad (8\text{-}2)$$

where n_s is the number of surviving units and N_o is the number originally placed in service. For the rank that drives the output rank, the population is reduced by the power gain or fan-out ratio $[G]$ and the probability of failure is reduced by the same factor. A failure in this rank reduces the drive power in a similar way and the loss in output is equal to the sum of the losses in dB calculated from the above equation. This assumes that the amplifier is linear. In practice, it is likely that the power is sufficient to drive the output rank into some degee of saturation, so that a dB loss of drive power will reduce output by only a fraction of a dB. The assumption of linearity will produce a conservative prediction of the transmitter's reliability.

8.1.1 Example of a CSA Reliability Prediction

A transmitter is constructed using RF amplifier modules that provide a power gain of 18 dB and produce a peak output power of 1600W. A reliability prediction has been performed on the module and it is expected to have a failure rate of $20/10^6$ hours when operated in the intended environment. The transmitter consists of a single module that drives 64 modules via a 1 to 64 power divider. The output of the 64 modules is fed to a 64 to 1 power combiner that delivers 92kW peak power to a matched load. The transmitter is required to deliver the rated power within one dB. Transmitter failure is declared when output power falls below 73kW. The power output loss is related to the number of surviving modules in the output rank by (8-2), from which we find $N_s = 57$ for a 1.0 dB loss. The equivalent failure rate for a group of 57 out of 64 units without replacement is found from [3]:

$$\lambda_e = \frac{\lambda_0}{\sum\limits_{i=A}^{N} (1/i)} \tag{8-3}$$

and

$$\lambda_e = \frac{\lambda_0}{\sum\limits_{57}^{64} (1/i)} = 7.55\lambda_0$$

It is now possible to construct a reliability block diagram of the transmitter as shown in Figure 8-2. It consists of two blocks in series, the first of which is the single driver module. The overall failure rate is the sum of the two block failure rates or 171 failures/10^6 hours. The corresponding MTBF is 5847 hours. The predicted operating life before the power output drops by one dB is approximately eight months. Maintenance could be scheduled for six-month intervals.

RELIABILITY BLOCK DIAGRAM OF SINGLE MODULE DRIVING A BANK OF 64 MODULES

Figure 8-2

The MTBF of the driver module is 50,000 hours. If it were backed up by a redundant spare with a failure rate assumed to be negligible until activated, the net impact on the transmitter MTBF is to increase it slightly to 6600 hours. The condition is more complex than it appears at first. A detector circuit must be designed to determine when the driver stage has failed and to switch on a spare module. The spare module must be tested periodically to ensure that it is ready to take over if needed. The reliability of the fault detection and switching circuit would have to be higher than that of the original driver module. If the fault detector had an MTBF of 100,000 hours and the switches had the same MTBF, the resulting MTBF of the fault detection and switching circuits would be the same as the original design without redundancy.

8.2 A SIMPLE RELIABILITY PREDICTION EQUATION

The preceding example illustrates the effect of a high module power gain on the overall transmitter reliability. The fan-out ratio is large, and most of the transistors are located in the output rank. If each module has the same failure rate λ, and there are N modules with power output P_j in the output rank, then the transmitter will average $N\lambda$ failures per hour and the output power will decrease after T hours by:

$$P = 2N\lambda TP_j \qquad (8\text{-}4)$$

The factor of two is an approximation (which is quite good for small fractional losses in the output rank) that accounts for the property of the output combiner that loss of a single module is equivalent to loss of twice the module output power. The power output loss after T hours of operation gives a degradation of:

$$D = \frac{10 \log(NP_j - 2N\lambda TP_j)}{NP_j}$$

or

$$D = 10 \log(1 - 2\lambda T), \quad dB \qquad (8\text{-}5)$$

This equation gives the approximate transmitter power loss after time T in terms of the module failure rate. For the previous example, the module failure rate was $20/10^6$ and the MTBF was found to be 5847 hours. Substituting the values in (8-4) gives a value of $D = -1.16$ dB, which is in good agreement with the 1 dB degradation employed in the reliability prediction.

8.3 RELIABILITY AND COST TRADE-OFFS BETWEEN SOLID-STATE AND VACUUM-TUBE TRANSMITTERS FOR PHASED ARRAY RADARS

The cost of ownership of a transmitter over a given useful equipment life consists of the sum of the costs of acquisition, operation, and maintenance. The acquisition cost consists of the costs of research and development, and the cost of manufacturing the transmitter. The manufacturing cost includes special tooling and test equipment. Operation and maintenance costs include the cost of operations and maintenance personnel, electric power, heat and cooling, logistical support, and the cost of spare parts including the cost of money (i.e., interest on loans, capital investment, depreciation, *et cetera*) required to provide the spares. A life-cycle cost comparison of a solid-state transmitter and a vacuum tube equivalent is developed in which an *N*-element phased array radar is to be operated for *L* years. Because of the complexity and time variability of cost modeling, this subject is deliberately avoided in favor of a cost comparison approach in which we assume a currently acceptable cost model has been applied to estimate the acquisition costs of both solid-state and vacuum-tube equivalent transmitter designs. The predicted failure rates and costs of replacement or repair of the tube and transceiver models are assumed to be known. The power efficiency of each type of transmitter is also assumed to be known. This information should allow a life-cycle cost analysis of the two alternatives to be performed so that a minimum cost system can be selected.

The vacuum tube transmitter powered phased array is shown schematically in Figure 8-3(a) and its solid-state counterpart is shown in Figure 8-3(b). The output of the tube transmitter is distributed to the array phase-shifters and radiating elements via a corporate feed network. The solid-state transmitter employs a similar network to feed RF drive power to the transceiver modules. The power loss associated with the drive power distribution is much lower than that of the tube transmitter because the transceiver drive power level is lower than the array-element power level by the power gain of the transceiver transmitting amplifier. With a 20–dB power gain module, a 2–dB loss in the drive power distribution network is equivalent to 0.07 dB loss in the output of the tube powered array feed network. Similarly, the phase shifter loss is also made negligible because of the transceiver module power gain. The phase shifter element is made less costly as well because it operates at a greatly reduced power level. The low-power level phase-shift elements may be produced economically using monolithic microwave integrated circuit (MMIC) technology. When the transceiver module is in the receive mode, a preamplifier, which establishes a low noise figure for the solid-state radar, provides a similar advantage over the tube powered system, which must pass the received signal through the phase shifters and corporate feed network before it traverses the duplexer and is fed to the receiver preamplifier.

The phase-shifter and power combiner losses are not critical in the solid-state system because the preamplifier precedes it. For a given radar detection-range performance requirement, the tube powered system must deliver a higher power output to overcome the transmission and reception losses in the phase shifter and corporate feed networks. The solid-state transceiver modules require a careful design, which must maintain a high degree of both phase and amplitude stability in transmit and receive modes.

PHASED ARRAY TRANSMITTER MODELS

(A)   TUBE TRANSMITTER

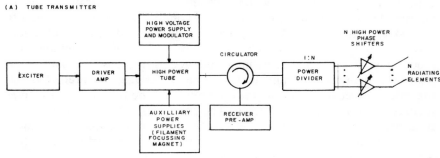

(B)   SOLID STATE TRANSMITTER

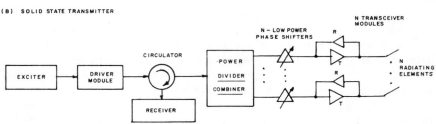

Figure 8-3

The life-cycle cost of the tube powered transmitter can be expressed as the following sum:

$$C_{tl} = C_{ta} + C_{to} + C_{tm} \tag{8-6}$$

where

C_{tl} = Tube transmitter life-cycle cost;
C_{ta} = Tube transmitter acquisition cost;
C_{to} = Tube transmitter operation cost;
C_{tm} = Tube transmitter maintenance cost.

Similarly, the life-cycle cost of the solid-state transmitter is

$$C_{sl} = C_{sa} + C_{so} + C_{sm} \tag{8-7}$$

where

C_{sl} = Solid-state life-cycle cost;
C_{sa} = Solid-state transmitter acquisition cost;
C_{so} = Solid-state transmitter operation cost;
C_{sm} = Solid-state transmitter maintenance cost.

The acquisition cost is a function of N, the number of elements in the array. The operating cost is related to the prime power consumed, or transmitter efficiency. The maintenance cost is a function of transmitter reliability. These factors are expanded below to derive the break-even cost trade-off relationships.

8.3.1 Acquisition Cost

The acquisition cost of both tube and solid-state transmitters will vary, not necessarily linearly, with the number of elements in the phased array. Because acquisition cost also depends on the manufacturing facility employed, it seems best to simply identify these cost elements as some function of the number of array elements. It is assumed that each user has an acquisition cost model which has been tested and found to be dependable. This type of data is usually proprietary.

The manufacturing cost typically drops over a period of time as the production crew learns to become more efficient and find short cuts to produce the product. This learning curve depends on the number of items produced and would take effect earlier in the production cycle of a solid state transmitter because the number of transceiver modules is generally equal to the number of array elements, while the tube transmitter employs only a small number of high-power rated components. The phase shifter elements of the vacuum-tube system would benefit from the large scale production.

8.3.2 Operating Cost

The cost of operation for each type of transmitter is primarily related to the cost of operations personnel, a cost which should not depend on whether the transmitter is a solid state design. A less significant cost, which does depend on the overall transmitter efficiency, is the cost of electric power to run the system. This cost is estimated from the power output and transmitter efficiency. The tube transmitter filament power supply, and losses associated with the phase shifters and feed network reduce the overall efficiency. In most cases, the difference in efficiency will not be a significant driver in the life cycle cost trade-off. For example, if a tube transmitter provides an average RF output power of 10kW with an efficiency of 35 percent, with output combiner and phase shifter losses of 2 dB, the net efficiency is reduced to 22 percent. If this were compared with a solid state transmitter with an overall efficiency of 35 percent, the difference in prime power consumed is approximately 17kW. At a prime power cost of 5 cents per kilowatt hour, the difference in annual operating cost is $7400, which is small compared to the other annual operating costs related to operation and maintenance. The annual fuel bill may be very important in a military system which must be operated at a remote site where the cost of supplies is very high. For situations where the cost of fuel is a significant fraction of the life cycle cost, the following may be included:

$$C_{so} = K(P_o/\eta_s)L$$

and

$$C_{to} = K(P_o/\alpha\eta_t)L$$

therefore,

$$C_{to} = C_{so}(\eta_s/\alpha\eta_t) \tag{8-8}$$

where

η_t = Tube transmitter efficiency;
P_o = RF output power, average kW;
η_s = Solid-state transmitter efficiency;
K = power cost ($/kWh) × 8760 = $/kW–yr;
α = Fraction of tube power output that reaches the array radiating elements;
L = Life-cycle in years.

8.3.3 Maintenance Cost

The cost of tube transmitter maintenance is equal to the cost of repair or replacement of a diverse set of transmitter components, which may include low-voltage circuit card assemblies in the low-power level stages as well as high-voltage rated transformers, capacitors, inductors, and microwave components. This includes the cost of training and supporting a maintenance crew, which may operate at great distances from the home office. The cost of spare components is a significant fraction of the maintenance cost. It must include the cost of spare parts and the organization and buildings necessary to store the parts, a management system that tracks spare parts utilization, a repair depot, and a transportation management system that ensures the timely delivery of vital field replacements when needed. For a tube powered transmitter, the maintenance cost is

$$C_{tm} = K_t Y L \lambda_t + C_{ta}(K_{ct}N_{ct}\lambda_{ct}L + K_{tl}) \tag{8-9}$$

where

K_t = Cost factor for repair of a failed tube expressed as a fraction of the cost of a new tube $(0 < K_t < 1.0)$;
K_t = 1.0 if the failed tube is discarded;
λ_t = Tube failure rate = 1/MTBF of tube;
λ_{ct} = Tube transmitter component failure rate (avg);
N_{ct} = Number of tube transmitter components;
Y = Cost of a new tube;
K_c = Fraction of the acquistion cost of the tube transmitter attributable to the random failure of a transmitter component $(0 < K_c < 1.0)$;
C_{ta} = Acquisition cost of the tube transmitter;
K_{tl} = Fraction of tube transmitter acquisition cost required to provide logistical support.

Similarly, for the solid-state transmitter, the corresponding maintenance cost is

$$C_{sm} = K_s MNL\lambda_m + MK_{cs}N_{cs}\lambda_{cs}L + C_{sa}K_{sl} \tag{8-10}$$

where
K_s = Fraction of the cost of a new transceiver module required to repair the average failed module. Only the transmitter portion of the module is considered $(0 < K_s < 1.0)$;
M = Module cost;
N = Number of array elements or modules;
λ_m = Module failure rate = 1/MTBF$_m$;
L = Operating life of system, years;
K_{sl} = Fraction of the solid-state transmitter acquisition cost required to provide logistical support;

N_{cs} = Number of components other than modules in the solid-state transmitter;

K_{cs} = Fraction of the module cost M attributable to a component failure in the solid-state system.

The component parts of a solid-state transmitter, other than the transceiver modules, include the low-voltage power supplies and regulators, the power and RF distribution networks, and the cooling and mechanical structure and connectors. It may be advantageous to include the dc regulator and power supply within the module package.

The component failure rates and cost factors shown above should be taken as a weighted average over the number of non-module components. This practice should also apply to the tube-transmitter component costs. A more accurate and rigorous approach would be to use the individual component failure rates and repair or replacement costs to arrive at the maintenance costs.

8.4 LIFE CYCLE COST TRADE-OFF ANALYSIS

The many complex elements of the life-cycle cost may be conveniently modeled with the aid of one of the many computer spread-sheet programs that are readily available for small personal computers. A hypothetical life-cycle cost spread-sheet analysis is presented here. The cost models given in the preceding sections are implemented to compare the ten-year life-cycle costs of a tube transmitter with an equivalent solid-state design. The acquisition cost of the solid-state transmitter is $5 million and the tube equivalent costs $4.5 million. The tube replacement cost $60,000, and a failed tube can be repaired for 75 percent of the cost of a new tube. The tube failure rate is estimated to be 1.43×10^{-4} failures per hour. The mean time between failures (MTBF) of the tube is approximately 7,000 hours. The tube is to power a 2000-element phased array with an average power of 10,000W. The loss in the RF distribution network and phase shifters is estimated to be 2.2 dB. There are 24 major components in the tube transmitter, which have an average failure rate of 5×10^{-5} (MTBF = 20,000 hours). The average cost of replacement of a failed tube transmitter is $4,500. The cost of power to run the transmitters is $438 per kW-year. A ten-year life-cycle is expected.

The solid-state transmitter module costs $2000 and failed modules are repaired for $200. The module MTBF might be as high as 100,000 hours, but lower reliability may be expected. The overall efficiency of the solid-state transmitter is 43 percent, the tube-transmitter efficiency is somewhat lower (because of the filament power supply and RF feed losses) at 35 percent.

The cost model spread-sheet is shown in Figure 8-4. The tube transmitter has a ten-year life cycle cost (LCC) of $9.8 million. The solid-state system has an LCC of $6.47 million with a module MTBF of 100,000 hours. As the module MTBF is reduced to 25,000 hours, the LCC increases to $7.52 million and the LCC difference between the two drops to $2.28 million. We may conclude there is ample margin for uncertainty regarding the module failure rate in this example. One reason the solid-state system does so well in this case is the relatively high logistical support cost of the tube system. The tube system must carry spares that cost 90 percent of the acquisition cost of the tube transmitter *versus* a 20 percent factor in the solid-state system. This somewhat oversimplified cost comparison illustrates the utility of the spread-sheet program for rapidly testing the sensitivity of the LCC to changes in the many cost elements. As was mentioned earlier, the cost of power in the two types of transmitters is small compared to costs of maintenance and logistics.

Figure 8-4

	RUN 1 MOD. MTBF=100000 HRS	RUN 2 MOD. MTBF=50000 HRS	RUN 3 MOD. MTBF=25000 HRS
Tube Transmitter Cost Factors			
K=Power Cost($/kWh)*8760	4.38E+02	4.38E+02	4.38E+02
Tube TX Efficiency	3.50E-01	3.50E-01	3.50E-01
Po=RF Power Output, kW	1.00E+01	1.00E+01	1.00E+01
Alpha=Power divider and Phasor Loss	6.00E-01	6.00E-01	6.00E-01
L=Life Cycle in years	1.00E+01	1.00E+01	1.00E+01
Kt=Tube repair cost factor	7.50E-01	7.50E-01	7.50E-01
Y=Cost of new tube	6.00E+04	6.00E+04	6.00E+04
Lambda(t)=Tube failure rate	1.43E-04	1.43E-04	1.43E-04
Lambda(ct)=Tube TX component failure rate	5.00E-05	5.00E-05	5.00E-05
Nct=Number of tube TX components	2.40E+01	2.40E+01	2.40E+01
Kct=Fraction of Tube Acq.Cost /failed component	1.00E-03	1.00E-03	1.00E-03
Ktl=Fraction of Tube TX for log.support	9.00E-01	9.00E-01	9.00E-01
Cto=Tube TX Operation Cost	2.09E+05	2.09E+05	2.09E+05
Ctm=Tube TX Maintenance Cost	5.09E+06	5.09E+06	5.09E+06
Cta=Tube TX Acquisition Cost	4.50E+06	4.50E+06	4.50E+06
Ctl=Tube TX Life Cycle Cost	9.80E+06	9.80E+06	9.80E+06
Solid State Transmitter Cost Factors			
Solid State TX Efficiency	4.30E-01	4.30E-01	4.30E-01
Ks=Fraction of module cost for module repair	1.00E-01	1.00E-01	1.00E-01
M=Transceiver Module Cost	2.00E+03	2.00E+03	2.00E+03
N=Number of modules	2.00E+03	2.00E+03	2.00E+03
Lambda(m)=Module failure rate	1.00E-05	1.00E-05	4.00E-05
Lambda(cs)=SSTX Component Failure Rate	1.00E-05	1.00E-05	1.00E-05
Ncs=Number of SSTX Components(not module)	4.00E+01	4.00E+01	4.00E+01
Kcs=Fraction of SS Module Cost/failed component	2.00E-01	2.00E-01	2.00E-01
Ksl=Logistic support cost factor	2.00E-01	2.00E-01	2.00E-01
Cso=Solid State TX Operation Cost	1.02E+05	1.02E+05	1.02E+05
Csm=Solid State TX Maintenance Cost	1.36E+06	1.71E+06	2.42E+06
Csa=Solid State TX Acquisition Cost	5.00E+06	5.00E+06	5.00E+06
Csl=Solid State TX Life Cycle Cost	6.47E+06	6.82E+06	7.52E+06
Csl-Ctl=LCC Difference	3.33E+06	2.98E+06	2.28E+06

Chapter 9

Examples of Successful Solid State Radar Designs

In this final chapter the design details of three successful designs are presented in the form of reprints of articles that have appeared in technical journals. The first of these is the now ten-year old AN/TPS-59, an L band tactical radar originally developed by the General Electric Corporation for the US Marine Corps. The TPS-59 is the first solid-state radar to make the transition from an experimental laboratory model to a full scale production product.

The first two articles are G.E. "white papers" on theTPS-59, then Bill Perkins of G.E. also gives a brief treatment of the other two radar transmitters selected for this chapter, the PAVE PAWS and the AN/SPS-40. The future of solid-state transmitters is also forecast in this article, which sees low cost phased array designs as possible through the development of microwave monolithic integrated circuit (MMIC) designs. The FET *versus* bipolar competition may also lead to higher frequency and higher efficiency designs.

A more detailed description of the AN/TPS-59 is given by E. J. Gersten and R. A. Joseph of G.E. The AN/TPS-59 is an L band tactical height-finder radar. Elevation beam-steering is provided by phase shifters in the elevation dimension, while the azimuth scanning is carried out by way of mechanical rotation of the antenna.

The next reprint is a *Microwave Journal* article by Donald Hoft, the project manager of the developmental model of the PAVE PAWS radar. This transmitter is also a first of a kind. It is the first solid-state transmitter powered high-power phased array radar. With 400 W per transceiver module and two 1792-element arrays, the peak power output of this UHF radar is 700 kW per array face. The radar was the first design in which a solid-state transmitter was selected over a vacuum-tube transmitter because of a life-cycle cost analysis, and its reliability has demonstrated field performance that is remarkably close to the predicted reliability.

The last article also reprinted from the *Microwave Journal* describes a solid-state transmitter that is also a first of its kind. The AN/SPS-40 is a shipboard UHF surveillance radar with a fairly low (5%) duty cycle.

The recent development showed that the size, weight, and performance of an older tube-powered transmitter could be duplicated or improved by a solid-state design.

The Westinghouse development of a solid-state replacement for a tube transmitter with a poor reliability record is the first documented case in which a retrofit was successfully deployed.

THE AN/TPS-59

Introduction

In June 1972, the Electronic Systems Division of the General Electric Company contracted with the U.S. Marine Corps to build an Engineering Development Model of a long-range, 3-D, L-band, air defense radar called the AN/TPS-59. It is unique in the sense that its RF electronics -- even its transmitter -- are totally solid-state. Its performance, notably in the areas of detectability in clutter, low-angle height finding, reliability, and maintainability, is substantially better than that of any other tactical radar.

Figure 1. AN/TPS-59 Radar System

Shown in the photograph, Figure 1, the radar antenna is a 9.1-m-high by 4.7-m-wide rotating planar array standing some 2.1 m above ground level. The three trailer assemblies which carry the array during transport are brought together in a "Y" configuration to support it during operation. The display and processing equipment, housed in two shelters, includes an AN/UYK-7 general-purpose computer.

The primary operational mode of the AN/TPS-59 is as a sensor, providing real-time, automatic, tricoordinate data on all targets within its surveillance volume to the Marine Corps' tactical data system. In an alternate, autonomous mode, it provides console readouts and controls to support ground-controlled intercept and air traffic control missions.

Performance Overview

Salient AN/TPS-59 performance characteristics are summarized in Table 1. Several of these characteristics are worthy of amplification. The elevation accuracy of 1.7 mrad translates to a height accuracy of 900 m at 540 km, decreasing linearly with range to 300 m at 180 km. Through implementation of a special low-angle height-finding technique, this accuracy is maintained to target elevation angles down to 0.8°, and a 2-mrad accuracy is achieved at 0.5°.

Parameter	Performance Level	Notes
Frequency	1215-1400 MHz	With pulse-to-pulse frequency agility
Surveillance Volume	360° azimuth 6-540 km range 0-19° elevation 10,000 ft altitude	Elevation coverage may be shifted electronically in 0.25° steps
Frame Time	10 or 5 seconds	
Detectability	P_D = 0.9 within 360 km = 0.7 from 360-450 km FAR = 5 per scan	Against a 1-m^2 fluctuating target
Resolution	Range = 61 m Azimuth = 3.2° Elevation = 1.6°	
Accuracy	Range = 24 m Azimuth = 3 mrad Elevation = 1.7 mrad	The approximately 17 to 1 beam-splitting is achieved with mono-pulse processing
Reliability	MTBF = 1000 hours	Calculation excludes IFF, prime power generation and shelter air conditioning and assumes a 600-hour preventive maintenance cycle
Maintainability	MTTR = 40 minutes	Maintenance performed by a two-man team
Weight	System = 13,600 kg Max Package = 2900 kg	
Prime Power Utilization	89 kW	Includes 23 kW for shelter air conditioning
Mobility	Assembly in 1 hour Disassembly in 1/2 hour	In daylight hours by a trained crew of 10 men
Transportability	Transport (C-130) Helicopter Truck (M35)	

The indicated detectability is maintained even with aircraft targets flying through a background of heavy ground clutter and/or heavy rain. Achieving this level of performance against clutter requires the following elements:

1) 53 dB signal-to-ground-clutter ratio improvement and 33 dB signal-to-weather-clutter ratio improvement, both achieved simultaneously.

2) Measurement of the mean weather doppler within each azimuth cell and retention of this information in the radar control computer.

3) Reconfiguration of the digital MTI on an azimuth cell-by-cell basis, under computer control, to implement a filter function with a clutter notch properly matched to the prevailing clutter spectrum.

4) MTI waveforms with pulse-to-pulse stagger to eliminate blind speeds.

5) L-band operation to keep the ratio of the weather spectral extent to the PRF to a manageable value. At S-band, with a PRF corresponding to a range of at least 60 km, the weather spectral extent is an unmanageably large fraction of the PRF.

The very high reliability of 1000 hours is due, in large measure, to the intrinsic redundancy of the solid-state, distributed RF electronics and the multichannel, primarily digital, signal processor. The major benefit of intrinsic redundancy is gradual, as opposed to catastrophic, failure. To take advantage of this characteristic, the AN/TPS-59 is designed with a sensitivity which is 2 dB in excess of that required to achieve specified performance. The radar will operate for a long period (1000 hours on the average) before it degrades in sensitivity by the 2-dB margin.

Antenna Design

The AN/TPS-59 antenna, illustrated in Figure 2, is made up of 54 row feed networks, associated row electronics, and row transmitter power supplies. The row electronics, composed of transmitter, preamplifiers, phase shifters, duplexer, and logic control, is housed in a single package directly behind the row feed. The vertical, or row-to-row, distribution is accomplished by three column feeds. Also located at the antenna, seated on the rotating platform, are the RF exciter, final receiver, and array control circuitry. Radar signals pass through the rotating joint at a 75-MHz IF, not at L-band.

The block diagram, Figure 3, describes the array operation. The RF exciter output is uniformly distributed by the transmit column feed to each row. It is phase shifted, amplified, filtered, and radiated by the row feed network. Received sum and azimuth difference signals are filtered and preamplified before phase shifting and summing in the column feeds. In addition to the column feed networks for collimating sum and azimuth difference signals, a third column feed network is used for receiving at low elevations and distributing exciter signals to the rows on transmit.

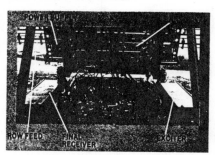

Figure 2.   Rear View of Array

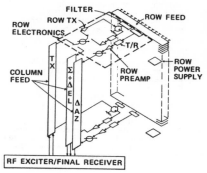

Figure 3.   Antenna Array Block Diagram

Implementing the array electronics on a per-row basis permits the vertical or row-to-row distribution and phase shifting to be performed at low level with little impact on system noise temperature --i.e., prior to high-power amplification on transmit and after preamplification on receive.

The physical proximity between row transmitter ouptput and radiating elements, and between receiving elements and row preamplifiers, minimizes losses and system noise figure. This translates to a reduction in required prime power of 3 dB or more relative to a more conventional tube radar performing a comparable mission.

Figure 4 shows one row feed network in a near-folded condition. It is a light-weight, single-layer strip line circuit etched on Teflon-glass and centered between ground planes by dielectric foam. A pair of hybrid couplers at the bottom center provide sum and azimuth difference outputs. An RF hinge permits the row-wing sections of the feed to be folded for tٍansport without breaking RF connections.

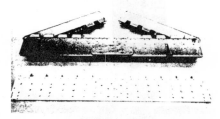

Figure 4.
Row Feed Network in Near-Folded Condition

As shown in Figure 5, ground plane septa are supported by brackets at each row. The septa provide an electrically solid surface that is mechanically open to reduce wind resistance.

The row electronics assembly is shown in Figure 6. The upper level houses all the low-power RF circuitry. Twenty-two RF power modules, which make up the row transmitter, are mounted on the upper surface of three bottom covers. The covers are finned to facilitate heat removal by forced convection using ambient air.

Figure 5.   Row Feed With Ground Plane Septa

The power module shown in Figure 7 in three stages of fabrication is the building block and replaceable unit of the solid-state transmitter. The power module operates as a Class-C amplifier, accepting an L-band input of 3 watts to output 50 watts. It contains three active devices--a driver stage and two parallel output power transistors. The module is enclosed on top with a plastic cover. The dc contact is located on the top of the module, and the RF contacts are located on opposite edges.

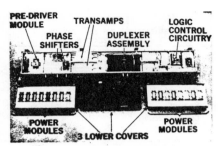

Figure 6. Row Electronics Circuitry

The power module operates at a dc level of 28 volts. This minimizes hazard to personnel and arcing problems. In fact, there will never be any waveguide arcing experienced on the AN/TPS-59 because it has no waveguide.

The schematic diagram, Figure 8, shows the electrical interconnection of the power modules within a row transmitter. The only impact of a power module failure is a reduction in the row transmitter output of 0.5 dB.

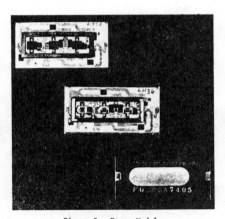

Figure 7. Power Module

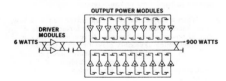

Figure 8.
Power Modules Within a Row Transmitter

Shelter Electronics

The signal and data processors and display console are housed in two shelters - the radar control shelter and the signal processor shelter.

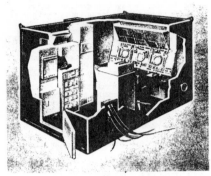

Figure 9. Radar Control Shelter

The radar control shelter is shown in Figure 9. All controls and indicators required for operation of the radar are located at the display console. As illustrated in Figure 10, two operator positions employing three cathode ray tube indicators are provided. These are the only vacuum tubes in the entire radar system. The outboard CRT's are PPI's. The central CRT presents an RHI display. The console includes IFF controls, a performance monitoring status display, a clutter gating panel, and communications controls.

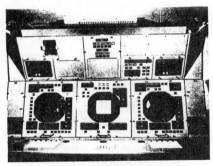

Figure 10. Display Console

The data processing subsystem consists of the AN/UYK-7 computer, its peripherals, and five hardware controllers which interface the computer to the external radar hardware. Peripherals include a magnetic tape system, which provides program storage, and the teletype/printer, which is the primary man/machine interface.

All aspects of radar operation are controlled by the AN/UYK-7. This control is a key factor underlying the overall performance capability of the radar, and it ensures the adaptability to a variety of new and different missions.

The signal processor shelter contains a double cabinet housing the waveform generator and pre-processor, primarily analog circuitry, and a three-bay cabinet with small digital boards which perform the bulk of the signal processing function. There are 3400 small printed circuit boards in the radar, of which 3100 are digital. However, among the 3100 are just 43 unique types, thereby minimizing sparing requirements.

As indicated in the block diagram, Figure 11, the signal processor, handling a variety of wave-forms, on up to four frequencies on each of four beams, is truly a multiwaveform, multichannel processor. The digital "integrate-and-dump" type MTI's cancel both ground and weather clutter on each of the four beams. Accepting 9-bit inputs, they will set a notch on ground clutter and cancel it to 56 dB.

A single LFM pulse compressor (BT 256) with 23-dB integration gain and -28-dB range sidelobes, operating at a 10-MHz rate, is multiplexed to simultaneously process up to 16 data channels.

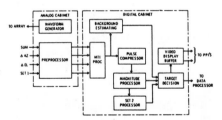

Figure 11. Signal Processor Block Diagram

Reliability, Maintainability, and Availability

The AN/TPS-59 will achieve full performance at 540 km until its sensitivity falls more than 2 dB below the design condition. Table 2 indicates one way in which this 2 dB may be distributed across the array. In any real situation, actual hardware losses will be distributed differently. However, this allocation does provide a basis for understanding the impact of the solid-state, distributed array electronics on reliability and availability. Note that the array will operate continuously for 4000 hours (5-1/2 months) without any preventive maintenance before this accumulation of failures may be expected.

TABLE 2

EXPECTED FAILURES FOR 4000 CONTINUOUS HOURS OF OPERATION

Function-Related Hardware	LRU Failures	Degradation (dB)
Power Module	13	0.104
Transmitter	8	1.392
Receiver	2	0.164
Common Receiver/ Transmitter	2	0.492
		2.152

Maintenance of the AN/TPS-59 is simplified by the automatic on-line performance monitoring (PM) and off-line fault location (FL) procedures. The PM process operates continuously, under computer control, using status messages and special loop tests scheduled between operational pulse repetition periods to evaluate all aspects of performance. A total of 469 special PM tests are run at a 6% duty factor to provide a thorough assessment of performance every 30 seconds.

The performance state of the system is indicated by the Performance Monitoring Status Display. This display, shown in Figure 12, contains nine lights, five related to hardware and four to performance. Normally, all indicators are on green. In the event of a failure, the appropriate light changes to either amber or red-amber if the system is in a partially degraded condition and red if it is seriously degraded.

Quantitative data accumulated in PM is presented, upon operator request, at the teletype/printer. Tables 3 and 4, listing the hardware performance data and computer program performance data available for output, indicate the comprehensiveness of the data. This data is all available with the system on-line.

Having determined that sufficient degradation exists to warrant shutting down for repair, the FL program is run after input to the computer from magnetic tape. It performs a large number of internal loop tests which indicate the equipment area in which the fault occurred.

The fault area size is not fixed. In the digital equipment, it is an average of 15 boards; in the analog equipment, it is generally one or two boards; and in the array electronics area, it is a particular row transmitter, row power supply, or row electronics package. The particular fault area is indicated in hard copy on the teletype/printer.

When a particular row transmitter output is low, power module fault isolation is performed using a special power module tester. It operates with the row electronics assembly closed so that the bad module is isolated before any modules are exposed. If a preamplifier, circulator, or duplexer in the row electronics fails, the row electronics assembly is replaced.

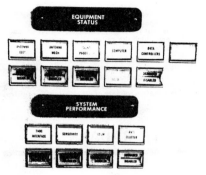

Figure 12. Performance Monitor Status Display

TABLE 3

HARDWARE PERFORMANCE DATA OUTPUT AT THE [1] *TELETYPE/PRINTER*

- Signal Processor Gain in Five Different Modes
- Angle Estimate Bias Errors With and Without STC
- ECCM Processing Gain
- Transmitter Output Power
- Individual Row Transmitter Output Levels
- Transmit Gain Loss
- Array Receive Gain
- Overall System Sensitivity
- Array Tilt and Temperature

TABLE 4

COMPUTER PROGRAM PERFORMANCE DATA OUTPUT AT THE TELETYPE/PRINTER

- Number of Target Reports Available to Data System Per Scan
- Number of Targets in Radar Storage Update Queue Per Scan
- Number of IFF Targets Processed Per Scan
- Number of Radar Targets Input Per Scan
- Number of Slots in Single Scan Store Released Per Scan
- Number of Azimuth Cells Adapted to Weather
- Number of Targets Not Reported to Data System Per Scan

Maintenance actions in the shelter are few and simple. Since items are replaced at the board level, only a printer circuit board tester, board pullers, and extenders are required. A maintenance manual is required in only the approximately 5% of the maintenance actions where automatic and semi-automatic procedures are not provided.

As a consequence of the high system reliability and the automatic nature of the PM and FL procedures, on-site maintenance may be accomplished with a two-man team. One of the maintenance men should be capable of repairing failures by remove-and-replace techniques (equivalent to a USAF Skill Level 3). The other maintenance man should be a technician capable of using an oscilloscope, power meters, and other standard test equipment.

System Testing

The AN/TPS-59 has been undergoing in-the-field testing since the beginning of 1975 in the harsh central New York State environment. It has experienced temperature extremes of -30°C to +35°C, heavy snow (305 cm per year), a severe ice storm, and a good deal of rain (132 cm per year).

During this interval the radar has successfully performed pattern tests, EMI and noise tests, and a heat test. Its detectability, in-the-clear and in clutter, are substantiated by more than 30 hours of controlled flight testing. Its reliability, maintainability, and, particularly, its availability have been substantiated by formal tests and by more than 18 months of 12-hour-a-day operation during which only one scheduled test had to be postponed because of a radar failure. Actual experience has clearly demonstrated that the AN/TPS-59 degrades gradually, not catastrophically. If it is needed for a mission, it will perform.

Having completed the contractor-run Acceptance Testing, the radar has been shipped to Camp Pendleton in California. After integration into the Marine Corps Tactical Data System and a short period devoted to hands-on training, it will begin a nine-month Marine Corps Service Test.

Modularity

An important quality of the AN/TPS-59 is its modular structure. This quality ensures that it can easily be reconfigured in response to a variety of threats and missions. For example:

- The array is made up of 54 identical row feed networks and associated row-level transceivers. These standard modules, or building blocks, used in varying quantities, will provide a wide range of power-aperture values.

- The AN/TPS-59 digital signal processor may be reconfigured to process up to 16 channels of data at bandwidths to 2.5 MHz. It match-filters either CW pulses or linearly frequency modulated signals with BT products of 64, 128, or 256. The digital MTI may be adjusted in real time under computer control to process any number of pulses with any set of complex weights and stagger codes.

Controlled by a general-purpose computer, many of the radar's operational parameters may be tailored to varying requirements through software modification alone. For example, changes may readily be made to

the waveform template, beam template, maximum range versus beam position, transmit frequency schedule, threat parameters which the radar will adapt to, and output data format.

The outstanding performance of the AN/TPS-59 and the modular nature of its design have prompted the conceptual design of a family of derivative solid-state radars, reconfigurations of the AN/TPS-59 which satisfy a variety of other mission profiles.

A FAMILY OF SOLID-STATE RADARS

3-D Long-Range Fixed Radar (GE592)

The GE592 is a fixed-site version of the AN/TPS-59. Figure 13 shows the antenna within a standard 55-foot radome and the Processing Center nearby. In this configuration the 7.3-m-square array is made up of 44 rows instead of 54, and the row feed networks are 50% longer than their AN/TPS-59 counterparts, but the row transceivers are the same as those employed on the AN/TPS-59.

Only low-level radar signals at IF, ac prime power, and video control signals are transmitted between the array and the Processing Center. Cables up to 4 km in length can be driven, so, as a practical alternative, the array may be operated remotely, up to 4 km away from the Processing Center.

The array may also be operated unattended. Its performance is automatically and continuously monitored at the Processing Center to ensure corrective maintenance when required. Array reliability is high enough to ensure proper operation with maintenance actions at intervals of not less than one to two months.

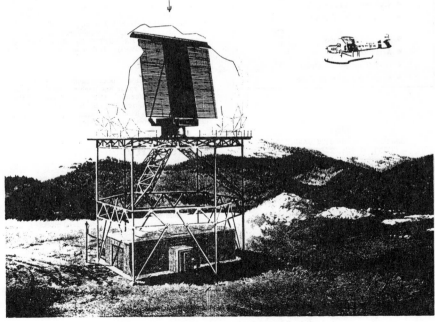

Figure 13. 3-D Long-Range Fixed Radar (GE592)

3-D Medium-Range Transportable Radar

The radar illustrated in Figure 14 is a medium-range version of the AN/TPS-59. The array is basically the lower half of the AN/TPS-59 array. It may be transported and supported during operation by one trailer. Also, the reduced range coverage and accuracy translates to less processing circuitry, which can fit comfortably in a single standard S-280 shelter. The resulting system is highly mobile, with a set-up time of 30 minutes. A summary of the performance characteristics of this radar is provided in Table 5.

A feature of this radar is that the general-purpose radar control computer, along with the console readouts and controls, will provide operator-designated tracking (on up to 50 tracks) and a computer-aided ground-controlled intercept capability (on up to 10 GCI's).

TABLE 5
PERFORMANCE CHARACTERISTICS OF 3-D MEDIUM-RANGE RADAR

Coverage	Range	225 km
	Azimuth	360°
	Altitude	100 kft
Resolution	Range	235 m
	Azimuth	3.2°
	Elevation	3.6°
Accuracy	Range	25 m
	Azimuth	5 mrad
	Height	3000 feet
Single-Scan Detectability	P_D	90%
	Cross-Section	1 m²
	P_{FA}	10^{-6}
Clutter Suppression	Ground	52 dB
	Weather	33 dB
R, M, A	MTBF	1000 hours
	MTTR	60 minutes
	Down Time/ Year	9 hours
Weight	System	4800 kg
	Max. Package	2500 kg
Prime Power		25 kW

Figure 14. Medium-Range Transportable Radar

2-D Long-Range Fixed Radar

If the AN/TPS-59 is turned on its side so that row feeds and transceivers become column feeds and transceivers, and the row-feed pencil-beam illumination is changed to a column-feed CSC^2 illumination, the 2-D long-range fixed radar illustrated in Figure 15 results. The antenna, made up of 32 columns, is 3.7 m high by 6.1 m wide. While shown in a rotating configuration, the antenna can also be electronically scanned in azimuth, thereby achieving inertialess coverage over a positionable $90-120^0$ sector with an attendant increase in reliability. If the requirement is for 360^0 azimuth coverage, whereby mechanical rotation is required, the electronic azimuth scan will still be useful, providing a look-back capability. A key point here is that the basic building blocks of the radar--the transceivers, the digital signal processing hardware, and the general-purpose computer program--are essentially the same as in the AN/TPS-59.

The performance characteristics of this radar are provided in Table 6. Among the features incorporated in the design are:

- Track-while-scan processing
- Unattended operation capability
- Remote performance monitoring
- Standard voice grade communication links

TABLE 6

PERFORMANCE CHARACTERISTICS OF 2-D LONG-RANGE FIXED RADAR

Coverage	Range	370 km
	Azimuth	360°
	Altitude	120 kft
Resolution	Range	3 km
	Azimuth	2.5°
	Elevation	CSC^2 to 45°
Accuracy	Range	3 km
	Azimuth	2.5°
Detectability (3 out of 4 scans)	P_D	99%
	Cross-Section	1 m^2
	P_{FA}	10^{-7}
Clutter Suppression	Ground	52 dB
R, M, A	MTBF	2400 hours
	MTTR	30 minutes
	Down Time/ Year	4.5 hours
Prime Power		25 kW

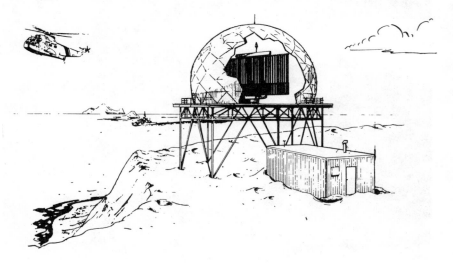

Figure 15. 2-D Long-Range Fixed Radar

2-D Medium-Range, Limited-Sector Radar

This radar, shown in Figure 16, is a smaller, shorter-range version of the long-range 2-D fixed radar. The antenna, made up of 12 columns, is 3.7 m high by 1.5 m wide. It is positionable to cover any selected 90-120° sector without rotating. It has basically the same performance as the long-range 2-D fixed radar, except that its detection range is 55 km, and, because it contains less processing equipment, it has a higher projected MTBF of 4000 hours.

Figure 16.   2-D Medium-Range Radar

AN/TPS-59 PERFORMANCE EVALUATION

System Description

The AN/TPS-59 is an L-band, all solid-state, 3-D surveillance radar developed for the Marine Corps.

The radar range coverage is from 3 to 300 nautical miles. It is scanned mechanically in azimuth through 360° and electrically in elevation from 0.5° to 19°. The maximum reporting height is 100,000 feet. The radar is designed to be trailer transportable, and consists of an antenna mounted on three trailers and two radar shelters (see Figure 1). One shelter houses the digital signal processor, waveform generator, and preprocessor while the other contains an AN/UYK-7 computer, peripherals, and three display consoles.

Figure 1. AN/TPS-59

The RF components relating to the transmit and receive functions are contained within the antenna structure in the form of solid-state row transceivers. There are 54 row transceivers, each driving a stripline board that feeds 28 dipole elements. Monopulse techniques in both azimuth and elevation are used to estimate target position. The overall dimensions of the array are 15 feet by 30 feet.

The radar has a time/energy management scheme that takes advantage of the inherent properties of solid-state transmitters – i.e., low peak power and high duty cycle. In the region from 3 to 20 nmi, a short simple pulse is used. From 20 to 100 nmi, a short-range LFM pulse is employed. Two long-range LFM pulses are used in the 100 to 200 nmi region, three pulses in the 200-250 nmi region, and four pulses in the 250-300 nmi region. The average duty cycle is 18%, with a peak radiated power of about 21 kW.

Parameters Tested

The test program of the AN/TPS-59 prior to delivery to the USMC included system integration and checkout, antenna pattern tests, a limited amount of environmental testing, radiation hazards determination, flight testing, and a reliability demonstration.

The parameters discussed in this paper include antenna patterns, flight test results (probability-of-detection; range, azimuth and height accuracies; and performance in clutter), and the results of the reliability demonstration.

Antenna Patterns

Patterns were measured at the GE Antenna Test Facility at Cazompy, N.Y. Principal plane cuts were made in elevation for both transmit and receive, and in azimuth in receive only. Patterns were taken as a function of frequency for both the sum and delta beams.

Azimuth Patterns

The aperture illumination in azimuth is predetermined by the row feed structure. An amplitude taper provides a Taylor 29 dB $\bar{n}$ = 5 weighting which is applicable to both transmit and receive sum. The delta azimuth weighting is a 22.5 dB $\bar{n}$ = 5 Bayliss.

The test method consisted of providing a CW L-band source in the far-field, and measuring the response of the antenna as a function of angular position. The array was raised to the perpendicular, in order to obtain principal axis cuts, by special backstay extenders. Patterns were taken using the antenna pedestal, rather than a separate mount. Servo motors were mounted in the azimuth drive train in order to provide an angular readout for the chart recorder.

A typical sum and delta pattern is shown in Figure 2. Beamwidths and sidelobe levels were comparable with predicted performance.

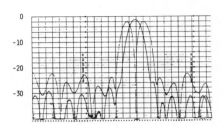

Figure 2. Azimuth Sum and Difference Pattern

Elevation Patterns

The aperture illumination in elevation is predetermined in amplitude by the taper in the column feeds. There is a uniformly weighted feed structure for transmit, a Taylor-weighted feed for receive sum, and a separate feed for the delta elevation beam. The column feeds are series-fed stripline devices.

To provide electronic scanning in elevation, there are phase shifters in each row transceiver. These phase shifters also remove any row-to-row phase mismatch in the aperture illumination.

Two types of patterns were measured, static and dynamic. Static patterns were taken by scanning the array mechanically in elevation. Dynamic patterns utilized the phase shifters to provide electronic scanning of the beam. A special computer program was required to provide the dynamic pattern capability.

Figure 3 is a receive sum static pattern, and Figure 4 shows the same beam in a dynamic pattern. The striking difference in sidelobe structure is due to the effects of the quantization errors in the illumination function resulting from the 4-bit phase shifters. As expected, this effect is quite pronounced because of the relatively few number of rows involved (54).

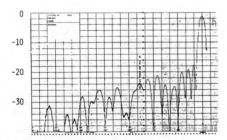

Figure 3.   Elevation Sum Pattern (Mechanical Scan)

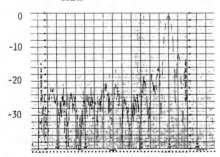

Figure 4.   Elevation Sum Pattern (Electronic Scan)

Flight Tests

The purpose of the flight testing was to provide the data necessary to evaluate the probability of detection; height, range, and azimuth accuracies; and performance in clutter.

An FPS-16 was used to accurately establish the position of the aircraft during the flights. The aircraft was a Learjet, and was equipped with a C-band beacon to improve the accuracy of the FPS-16 estimates.

The tactical software was modified to provide a magnetic tape recording option. Two recording modes were established. One allowed the data on a single target to be recorded every scan, while the other permitted the recording of data on all targets detected during each scan. The former was used for collecting data on the controlled aircraft, and the latter was used to obtain false alarm statistics.

In addition to tactical software modification, special hardware was added to allow synchronous interrogation of the tracking radar with the TPS-59. Special software was also required to record the FPS-16 data.

The flight paths were radial legs of altitudes ranging from 2500 ft to 40,000 ft, and at ranges out to 200 nmi (i.e., at ranges out to radar horizon). Flights were also run through regions of heavy ground clutter in order to evaluate MTI performance. The data from the '59 and the '16 was reduced off-line. In all cases, the difference between the '59 position estimate and the '16 estimate was considered a '59 error.

Predictions of performance prior to the start of the flight tests were made difficult because of the unknown character of the target cross-section. However, accurate determinations of system sensitivity were made. Based on the assumption that the aircraft in a nose-on or tail-on aspect would represent a 0.7 m^2 non-fluctuating target, it was predicted that we would be able to detect it out to 200 nmi with a P_D greater than 0.9, even if the system was degraded by 2 dB.

The results of the testing are shown in Figure 5. This represents the performance averaged across all flights as a function of range. During the flights, the radar was degraded by 0.5 dB due to transmit rows being out. There is an obvious discrepancy in the data in the region from 80-100 nmi and 180-200 nmi.

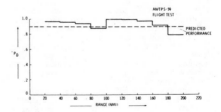

Figure 5.   Probability of Detection Versus Range

Analysis of the results, taking into account the difference in transmit template at the two ranges, indicated that the target model that would be consistent with the observations would be a 0.7 m^2 Swirling Case III. This appears to be consistent with the constant aspect angle that the airplane assumed - i.e., a dominant component surrounded by a lot of small fluctuating components.

The range and azimuth accuracies are shown in Figures 6 and 7. These accuracies are defined as the standard deviation of the position error between the '59 and '16. The curves represent the summation of all flights.

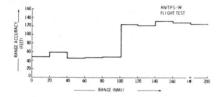

Figure 6. Range Accuracy Versus Range

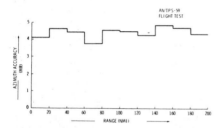

Figure 7. Azimuth Accuracy Versus Range

The jump in range accuracy at 100 nmi is due to the change in the transmit waveform template at that point. The interesting thing about the azimuth accuracy is that it doesn't vary with the changes in signal- to-noise. This implies that the major component of the azimuth error is not related to system sensitivity. Subsequent investigations showed that the azimuth encoder was excessively noisy. The encoder has been repaired and reinstalled on the equipment at MCTSSA.

The height accuracy is shown in Figure 8. Height accuracy for this case is defined differently from range and azimuth. Since a constant elevation angle bias will result in an increasing height error with range, the height bias is included in the accuracy calculations.

Analysis of the data indicated an elevation bias of 1.5 mr. This was removed via a simple software change prior to delivery of the radar to the Marine Corps. This will reduce the height error by approximately 30% at the longer ranges.

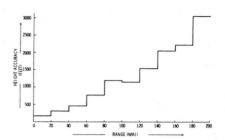

Figure 8. Height Accuracy Versus Range

To evaluate the performance of the radar in clutter, several flights were run through heavy clutter regions. The speed of the aircraft was changed from run to run in order to demonstrate the absence of blind speeds in the MTI circuitry. The parameter of interest was probability of detection, with the ground rule that a detection in any pair of scans would count as a detection.

Under these conditions, the probability of detection was essentially unity from 20 to 80 nmi.

Reliability Demonstration

One of the unique features of the AN/TPS-59 is its all solid-state construction, and hence inherent high reliability. It was very important to demonstrate this reliability as soon as it was practical to do so. Two weeks after flight tests, a 540-hour reliability demonstration was conducted.

The daily test schedule called for 45 minutes of "power off" time, a 15-minute period to check the status of the system, and 23 hours of normal tactical operation. The built-in performance monitoring features of the AN/TPS-59 allowed continuous availability of the status of the system.

The radar was designed with a 2 dB guardband, to allow a certain number of failures in the redundant areas of the equipment without a resultant system failure. Thus, component failures were not necessarily classified as a relevant failure of the system.

During the 540-hour run, there was a failure of a digital board in the processor, and eight row-transceiver-related failures. The failed digital board was in the MTI equipment, and although MTI performance was degraded, it did not constitute a system failure. The eight transceiver-related failures resulted in a sensitivity degradation of 1.4 dB versus the built-in field degradation guardband of 2 dB.

The projected MTBF, based on 540 hours with no failures, is 589 hours with 60% confidence.

Summary

The mechanical, electrical and system design features of the AN/TPS-59 provided certain advantages with respect to test methods and system evaluation.

The mechanical configuration of the antenna allowed patterns to be measured using both electrical and mechanical scanning techniques. No special mounts were required.

The tactical software required only a minor modification to permit data recording on magnetic tape for flight testing. The different pulse widths and transmit templates at the longer ranges facilitated the analysis of system performance.

There were several unique features of the reliability demonstration. The system design provided continuous on-line monitoring of the status of the radar. There was daily exercising of the fault location and isolation system. The test schedule introduced turn-on/off transients that are typical of tactical operations. During the 540-hour test interval, the equipment was subjected to a wide variety of environmental conditions

The net result of the test program was a clear, accurate picture of the performance characteristics of the AN/TPS-59 EDM radar.

AN/TPS-59 Performance Evaluation, by E. J. Gersten and R. A. Joseph, General Electric Company, Syracuse, N.Y., was originally presented at the EASCON '76 Symposium, September 1976.

Solid-State Devices for Radar
William H. Perkins
General Electric Co.

The impact of solid-state technology on radar is traced with special emphasis on RF power generation. The first solid-state phased array radars are reviewed. Recent solid-state monolithic microwave integrated circuit (MMIC) technology is reviewed to show that this new technology provides the promise of low cost microwave circuits, which should make phased array radars affordable on an initial cost basis. Advanced radar concepts then are discussed in light of the potential of low cost microwave ICs in the future.

Introduction

It is interesting to note that solid-state devices played an important role in the original development of radio. The earliest radio frequency detectors were metal-oxided coherers, and these were followed by point-contact semiconductor diodes using lead sulfide (galena crystals). Although the generation of radio frequency signals began with spark-gap transmitters, these were supplemented by the "solid-state" alternators of E.F.W. Alexanderson for high-power transatlantic communications during World War I. Lee DeForest's invention in 1907 of the "audion," a high vacuum triode, ushered in the era of vacuum tubes, which pushed most solid-state radio frequency devices into the background after World War I was over.

The development of radar began with vacuum triode and tetrode transmitters. During World War II, as frequencies were pushed into the microwave region, the magnetron became the dominant radar transmitting tube. Kylstron amplifiers, developed during World War II, led to improved high-power kylstron and traveling-wave tube amplifier transmitters since the war. Solid-state diodes, used as receiving mixers and detectors, were widely used in radar as higher microwave frequencies made tube performance less suitable.

Just as DeForest's invention of the three-element vacuum tube was an important turning point in the history of radar, the invention of a three-element solid-state device, the transistor, by Bardeen, Brattain, and Shockley at Bell Telephone Laboratories in 1948 was a very significant turning point, away from vacuum tubes. The transition from point-contact transistors to junction transistors and from germanium to silicon in the 1950s began the development of higher power transistors, but their power was still small compared to the powers attained by transmitting vacuum tubes.

The concept of phased-array radar antennas in which the power was distributed a large number of radiating elements suggested that the transmitting amplifiers need not be very large. Texas Instruments began solid-state transmitter array development in late 1963 on the X-band MERA contract for the Air Force and in 1967 at L band on the ARPA/ABMDA Camel program, which also involved other contractors.

At frequencies in the range from S through X band, gallium arsenide field-effect transistors (GaAs FETs) and amplifiers have been developed which could provide suitable power for array radar application.

At present, the most important trend in solid-state radar transmitters is the development of monolithic microwave integrated circuits. These provide complete amplifiers on single GaAs chips up through X-band frequencies. In addition, the phase shifter and receiving components needed for complete radar array elemental transceivers have been built in chip form. The current status and future development of these monolithic circuits is briefly described in this article.

The last section of the article describes the configurations of future radars that will meet more stringent military requirements. The necessity for low cost transmit and receive modules is pointed out, along with the potential of monolithic microwave integrated circuits to meet this critical need.

Present Day Solid-State Radars

An accurate measure of the impact of solid-state devices on radar can be ascertained by considering several key radar examples: L-band tactical solid-state radars, the PAVE PAWS surveillance radar, and the AN/SPS-40 radar. Airborne phased array developments also will be briefly considered.

L-Band Solid-State Radars

Both the General Electric and Marconi companies now have radars that use L-band bipolar power transistors as means for RF power generation. Experience of the GE L-band family of radars is described, followed by a discussion of the Marconi radar.

The solid-state radars in the GE family include the AN/TPS-59 for the Marine Corps, AN/FPS-117 for the Air Force, and the GE592, which is being sold to NATO. Figure 1 shows the AN/TPS-59 version; it is designed to provide long range surveillance and ground control intercept capabilities within a range up to 300 nautical miles. A single array face is mechanically rotated to provide 360 degree coverage in azimuth in this type of radar. Coverage in elevation is achieved by electronic steering of the phase of each row of the radar. Therefore, identical solid-state phase shifters, row transmitters, and sum and difference receivers are used for each row of the array.

FIGURE 1

The key to the development of these L-band radars was the development of the microwave silicon bipolar transistor. Performance of the transistor determined the magnitude of several problems. Low transistor power required excessive combining to accomplish the required one kilowatt of peak power per row transmitter, low efficiency reduced the reliability of the array and required more expensive cooling techniques, and the many transistor amplifiers and connectors presented a major mechanical problem.

A brief review of the stages of the development of the row transmitters provides a look at the rapid development that took place for solid-state radar transmitters. In 1971 the best available bipolar transistor provided 12 watts peak power at L-band frequencies. Three of these transistors, one driving two output stages, were used in the development of a 20 watt module. Forty of these modules were combined to generate 750 watts peak power.

By 1973, 25 watt transistors were available and 50 watt modules were used in the first engineering development model (EDM) of the Marine Corps' AN/TPS-59. The performance of these transistors allowed the combining of 20 modules to achieve one kilowatt peak power. Efficiency of each module was 35 percent, which allowed the row transistors to be air-cooled and still achieve the reliability requirements.

By 1976 silicon L-band power transistors had achieved 55 watt CW power output with 55 percent efficiency. Thus, the module was again designed as a 100 watt module and only 10 modules were required in each row transistor. The efficiency achieved by the 100 watt modules was 40 to 45 percent, which provided a further improvement in reliability and mechanical asembly. The corresponding one kilowatt row transmitter efficiency was 33 to 38 percent, which allowed the tight prime power budget to be achieved.

Extensive development of module and row transistor package concepts took place throughout the design of the 25, 50, and 100 watt modules. Figure 2 shows the present 100 watt module as now made in production. The large area hermetic seal was achieved by using a ceramic wall and a copper frame directly bonded to the alumina substrate. The input and output RF and DC contacts are made via spring-loaded contacts to avoid the proliferation of microwave connectors and cables.

FIGURE 3

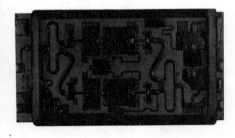

FIGURE 2

The transistors are mounted on beryllia carriers to enhance the heat transfer to a direct-bonded copper and molybdenum baseplate. The circuit shown in Figure 2 shows an input transistor and the two 55 watt transistors which are combined to provide the 100 watt output.

In the completed row transistor, the DC and RF circuits are housed in the lid of the assembly; contact to the modules occurs when the lid is closed. An additional key feature of this type of row transistor is the ability to vary the number of modules in each row transistor, thereby achieving amplitude taper across the array face. Another feature of the row transistor, which is really a feature of all solid-state arrays, is that a module failure means a very slight performance degradation. This feature has meant a very significant improvement in reliability compared to radars with central, tube type, high-power transmitters.

Marconi Solid-State Tactical Radar

Another L-band solid-state radar is made by the Marconi company. Called the S-723A and 723C, it is a variant of the tube-type radar (S-713). The principal difference between the S-723 and the S-713 is the use of multiple modular solid-state modules, which are distributed along the center of the array rather than the use of a central tube-type transmitter Each solid-state transmitter drives a row of ter consists of 16 identical, 150 watt, bipolar transistors operating in parallel. The 16 transistors are driven by a four-transistor driver amplier.[1]

Performance of the bipolar transistor is now adequate for array radars below S band. Transistor peak power performance presently exceeds 250 watts for short, less than 10 μs, pulse widths and duty factors of 5 percent.

One of the earliest uses of GaAs FETs was for low noise receivers. In this class of L-band radars, the transistors are used to establish the noise figure in the row receivers. PIN diode phase shifters are used to provide phase steering of the rows. The only non-semiconductor component throughout this class of radar is the display tube for manual operator tracking of targets.

Have solid-state devices verified the earlier expectations in terms of performance, reliability, and lower life cycle cost for this class of radar? The answer is yes. Recent reliability data for the GE592 radar is greater than 1,000 hours mean time before failure (MTBF) and requires less than 40 minutes for mean time to repair (MTTR). A 750,000 hours MTBF for individual 100 watt L-band power modules has been achieved.

 Likewise, the results on the S-723 solid-state radar also verify the superior reliability performance of solid-state radars. Two-hundred thousand hours of testing have shown no failures.[2] Both the GE and Marconi radars use bipolar power transistors that are supplied by several companies and that use mature fabrication technology.

PAVE PAWS

The only operational solid-state radar that utilizes transmit and receive solid-state modules at each radiating element is the Raytheon PAVE PAWS radar. This large radar, shown in Figure 4, also uses silicon bipolar power transistors to provide the necessary RF transmit signal. Thereare 3,584 active solid-state modules per radar.[3] These modules were configured as shown in the block diagram in Figure 5.

FIGURE 4

One of the most significant accomplishments was the efficient combining of the power transistors to generate the radiated power. Figure 6 shows the power amplifier configuration. An average of 322 watts of output power per module was achieved at UHF. The key to the success of this radar was the ability to manufacture modules that were sufficiently identical.

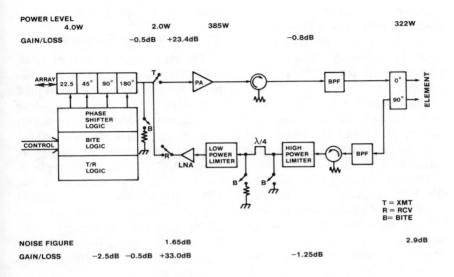

POWER LEVEL
4.0W   2.0W   385W   322W

GAIN/LOSS   −0.5dB   +23.4dB   −0.8dB

T = XMT
R = RCV
B= BITE

NOISE FIGURE   1.65dB   2.9dB

GAIN/LOSS   −2.5dB   −0.5dB   +33.0dB   −1.25dB

FIGURE 5. UHF SOLID-STATE MODULE BLOCK DIAGRAM.

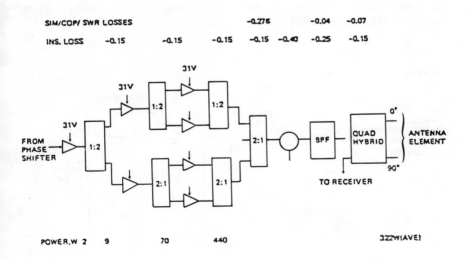

SIM/COP/ SWR LOSSES   −0.275   −0.04   −0.07

INS. LOSS   −0.15   −0.15   −0.15   −0.15   −0.40   −0.25   −0.15

POWER,W 2   9   70   440   322W(AVE)

FIGURE 6

Figure 7 shows the transmitter/ receiver configuration. Placing the modules near the radiating elements reduces the distribution losses by at least 3 dB over radars with centrally located tube transmitters.

The calculated reliability of the PAVE PAWS module was over 220,000 hours MTBF. The experience to date has exceeded these earlier estimates.

Data accumulated on both the GE L-band radars and the Raytheon PAVE PAWS systems verifies that solid-state radars have the advantages of reduced prime power, greater reliability, and lower rate of repair when compared to radars with tube-type transmitters. Graceful or gradual degradation of the radars avoids the catastrophic failures experienced with previous radars.

Planned, routine maintenance also greatly minimizes life cycle cost.

AN/SPS-40 Radar Transmitter

The most direct impact of solid-state components on radar has been the direct replacement of tube amplifiers with their solid-state equivalents. Solid-state upgrades to existing radars are now taking place at frequencies up through K band. An X-band TWT replacement now is in production for the DSCS III satellite communication upgrade.[4] This trend will continue as the performance improves and the cost of generating solid-state power continues to decrease.[5]

A key demonstration of the usefulness of solid-state transmitters has been the development and application of a 250 kW solid-state transmitter for the AN/SPS-40 radar.[6]

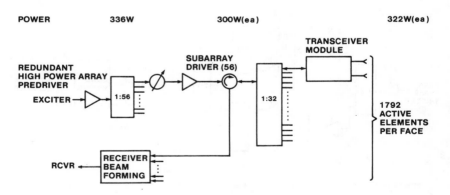

FIGURE 7. PAVE PAWS TRANSMITTER/RECEIVER CONFIGURATION.

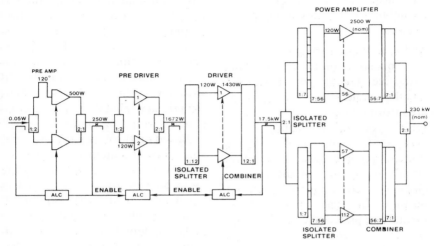

FIGURE 8. AN/SPS-40 TRANSMITTER BLOCK DIAGRAM.

The block diagram for the transmitter is shown in Figure 8. The output stage is made up of 112 identical 2,500 watt modules paralleled to generate the total 250 kilowatts of peak power. The 112 modules are housed in two cabinets. A third cabinet houses the predriver and driver circuits.

This solid-state transmitter met all of the performance requirements derived originally for the tube transmitter. In addition, it achieved variable output power, gradual power degradation over the life of the equipment, and higher efficiency. The key improvement, however, was reliability. A 10,000 hour mean time between failure (MTBF) is expected due to the built-in redundancy and conservative design.

100 KW L-Band Solid-State Transmitter Design

ITT Gilfillan has developed a 100 kW L-band transmitter design that takes advantage of bipolar transistor performance. This development was specifically for minimally attended radar (MAR) applications.[7] The transmitter consists of 110, 1100 watt modules being combined using a parallel plate 110-way combiner. Actual performance achieved with a smaller 20 module transmitter was 17 kW. This power output, coupled with good efficiency (31 percent and 61 dB gain, demonstrated that solid-state L-band transmitters were a vaiable alternative to the tube-type predecessors.

Airborne Phased Arrays

A question that has stimulated extensive development activity is whether solid-state phased arrays are affordable at frequencies above UHF. As the frequency is increased, the number of elemental transmit/receive modules required greatly increases and the cost per element becomes increasingly important. In 1968, Texas Instruments completed the MERA (molecular electronic radar applications) elemental radar. Following that development, TI developed the RASSAR (reliable advanced solid-state radar).[8]

Hughes is developing X-band mini-hybrid modules designed for producibility. Their emphasis is automation, and their intent is to stress automated assembly and test to reduce cost of solid-state hybrid modules for airborne radar systems.[9] By this means, Hughes expects to achieve affordable airborne phased array radars. If successful, the TWT transmitter, the primary reliability limiting component, would be eliminated.

Based on the experience to date, solid-state components have increased reliability and lowered life cycle cost of ground based radars when applied at frequencies of L band and lower. The challenge is to make elemental phased array solid-state radars affordable at higher frequencies.

Solid-State Technology Status for Radar Transmitters

Two principal semiconductor technologies are impacting radar design. As previously discussed,

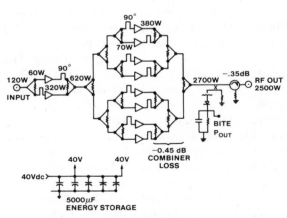

FIGURE 9. POWER AMPLIFIER MODULE BLOCK DIAGRAM.

silicon bipolar transistors now are in use in radars at L-band frequencies and below. Also, silicon bipolar monolithic integrated circuits are being sold commercially.[10] These new low-powered microwave integrated circuits are being made in large quantities at low cost due to high yield fabrications on 4- and 5-inch wafers.

The second semiconductor material, GAAs, is being used in both discrete and integrated circuit form to provide high performance microwave components. GaAs material has several features that provide improved microwave performance over silicon. The two materials are compared in Table 1.

Performance of transistors made with the two types of material is shown in Figure 10.[11-23] As can be seen, useful performance for silicon is well below 10 GHz, where the GaAs MESFET can produce useful gain at 60 GHz by using submicron gate geometries.[24] As yields improve, the cost of transmit/receive modules using discrete and monolithic GaAs components is sure to decrease.

GaAs MMICs

Some of the major attractions of the monolithic microwave integrated circuit (MMIC) technology, both in silicon and GaAs, that are now being pursued by radar companies and government agencies are:

TABLE I

	Si	GaAs
ENERGY GAP (eV) at 300°K	1.12	1.424
MINORITY CARRIER LIFETIME (SEC)	2×10^{-3}	$\sim 10^{-8}$
ELECTRON MOBILITY (DRIFT) $CM^2/V\text{-}S$ ($10^{17}/cm^3$)	700	5000
DRIFT VELOCITY (CM/SEC)	1×10^7	2×10^7
FOR $N_D = 10^{17}CM^{-3}$, T = 300°K		
THERMAL CONDUCTIVITY $W/CM^2\text{-}°C$) AT 300°K	1.5	0.46
TEMPERATURE COEFFICIENT OF EXPANSION ($\times 10^{-6}/°C$) AT 300°K	2.3	5.8
VAPOR PRESSURE (P_A)	1 AT 1650°C	100 AT 1050°C
RESISTIVITY (Ω-cm) AT 300°K (UNDOPED)	10^3	10^7

Primary Source: PHYSICS OF SEMICONDUCTORS DEVICES by S.M. Sze

COMPARISON OF SEMI-CONDUCTOR MATERIALS

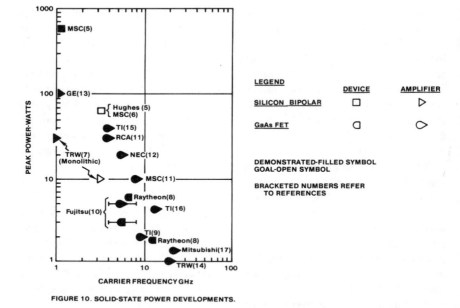

FIGURE 10. SOLID-STATE POWER DEVELOPMENTS.

- Potentially lower cost—wafer fabrication with greater than 2000 50 × 50 mil MMICs on a 3-inch wafer
- Higher reliability—reduction in transmit/receiver component count by a factor of 10 is feasible
- Circuit-to-circuit uniformity—elimination of manual tuning
- Superior performance—especially at frequencies above 10 GHz
- Smaller size and weight—dependent on the degree of on-chip integration, but reductions of 100 times smaller are feasible.

MMIC designs from VHF through Ka-band are now under development. As shown in Figure 11, the lower frequency MMICs require the use of lumped element components rather than distributed

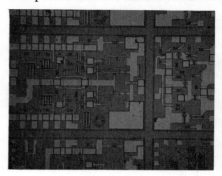

FIGURE 11

transmission lines in order to achieve a small chip size. However, at the higher frequencies (see Figure 12), distributed lines can also be used without seriously increasing the total chip size. The size of an L-band hybrid, two-stage low noise amplifier shown in Figure 13 is compared to an L-band, two-stage monolithic amplifier. The ease of tuning the hybrid stripline is performed by the tunable capacitors, while the difficulty of tuning an MMIC is evident.

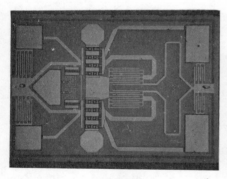

FIGURE 12

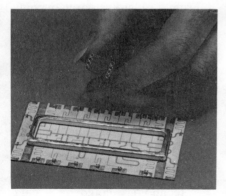

FIGURE 13

In Figure 14, a four-bit phase shifter MMIC is compared to an equivalent hybrid circuit built on an alumina substrate. Not only is there a potential of great size reduction, but also the uniformity achieved with wafer fabrication eliminates or greatly reduces the tuning of individual circuits. For a phased array radar, the transmit/receive modules must be nearly identical in order to achieve optimum performance. Thus, the MMIC technology holds the promise of reduced sidelobes due to superior phase and amplitude tracking of the modules located at each

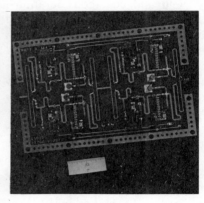

FIGURE 14

radiating element. This feature is especially critical at higher microwave frequencies where achievable dimensional tolerances are a much larger fraction of a wavelength.

Reduction in the number of components also is achieved with the MMIC technology. As shown in the previous figures, a single GaAs chip can take the place of a hybrid circuit, such as the four-bit phase shifter, which has as many as 50 individual components.

Although MMIC technology is not without its problems, most problem areas will be reduced as the technology matures. Rapid progress is being made in GaAs materials, CAD and wafer handling.

Yield improvements of both discrete and integrated microwave components will result in lower-cost, reproducible modules. The degree of on-chip integration vs. hybrid assembly will be determined by yield and cost considerations. Extensive efforts to automate and control performance variations, both at wafer fabrication and at module assembly and test levels, must be made to achieve the low cost modules necessary for higher frequency phased array radars.

As in silicon, the trend to increase the number of functions on a single chip will continue in GaAs ICs. GaAs, which has electron mobilities six times higher than silicon, will enable higher-speed GaAs digital interface ICs to be used with silicon VLSI in high speed processors. Also, the ability to incorporate both microwave and digital functions on the same chip will enable the microwave-to-digital interface to occur much closer to the RF sensor in elemental phased array radars.

Cost trends are a measure of the availability of GaAs MMICs for radars of the future. Based on the cost projection shown in Figure 15, developed by Ron Naster, MMIC Program manager of General Electric's Electronics Laboratory, MMIC cost will be reduced by a factor of at least four by 1989.

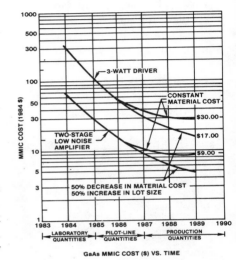

GaAs MMIC COST ($) VS. TIME

FIGURE 15

Impact of Solid-State on Future Radars

From a radar system viewpoint, the payoff for solid-state technology can be summarized in three general areas:

- Those areas where solid-state is essential both now and in the future;
- Those areas where solid-state technology could increase affordability of today's expensive phased array radars;
- Those areas where solid-state technology could be a catalyst for revolutionary new ideas in tomorrow's radars.

Solid-state components are necessary in radar applications where weight, volume, and reliability are key radar design drivers, such as airborne and satellite systems. Thus, solid-state technology should play a key role in the continuing development of airborne and space-based radars in particular and in future developments of minimally attended and unattended radars in general.

Today's phased array radars are monostatic radars, which means the transmit and receive antennas are colocated; moreover, today's radars usually are single-beam radars with single transmit and single receive beams with a common transmit/receive aperture.

Tomorrow's radars can use solid-state and extensive digital technology to address many of the shortcomings of today's radars and to provide capability against increasing threats and more demanding mission requirements.

What might the new radar capabilities be, and what is the enabling role of solid-state technology? Low cost transmit and receive modules are required for effective distributed radars, such as bistatic and multistatic radar systems. A bistatic radar system consists of a single transmitter and receiver that are spatially separated. A multistatic radar system consists of multiple spatially separated transmitters and receivers.

Survivability against anti-radiation weapons is one of the most fundamental requirements of future radars. Distributed radars with multiple transmitters enhance survivability over that of monostatic radars.

Distributed radars require multiple simultaneous high resolution receive beams so that they can efficiently collect the energy from broadened remotely located transmit beams. This requires low cost phased arrays which can be provided by the use of monolithic microwave integrated circuits (MMICs). Monolithic components will play a key role in future digital radars which utilize a monolithic receiver and A/D converter for each subarray. Following the A/D converters will be digital beamforming networks. This kind of array can adaptively steer multiple high performance receive beams over a large volume in space.

Both industry and government resources must be focused to achieve the low cost MMIC modules for the future. Cost leverage will not only allow us to make today's phased array radars more affordable, but also will open the way to more effective radar system concepts in the future.

References

1. "Marconi to Display Air Defense, Terminal Radar at Farnborough," *Aviation Week and Space Technology,* Vol. 121, No. 6, p. 81, August 6, 1984.

2. Payne, M., "Marconi Shows Off Martello Variant," *Electronics Weekly* (UK), No. 1227, p. 66, July 25, 1984.

3. Hoft, D. J., "Solid-State Transmit/Receive Module for PAVE PAWS Phased Array Radar," *Microwave Journal*, Vol. 21, No. 10, pp. 33–35, October, 1978.

4. SooHoo, J., and J. Hrinkevich, "Upgraded Systems Performance of DSCS III Satellite Using GaAs FET Solid-State Amplifiers," 1984 IEE MILCOM Conference.

5. Perkins, W. H., "Solid-State Cost Effective Radars," EASCON, 1975.

6. Lee, K., C. Corson, G. Mols, "A 250 KW Solid-State AN/TPS-40 Radar Transmitter," *Microwave Journal* pp. 93–105, July, 1983.

7. Sanders, B. J., "Solid-State Transmitters for Advanced Radar Systems," WESCON Profesional Program, September, 1980.

8. Klass, P. J., "New Array to Impede Radar Failures," *Aviation Week and Space Technology,* January 4, 1982.

9. Personal communication with Dr. Gene Gregory, program manager for module development at Hughes Aircraft in Torrance, CA.

10. Huang, L. and N. Osbrink, "Ku-Band LNC Offers High Performance With Its Silicon Monolithic IF Amplifier," *MSN*, pp. 46–52, July, 1983.

11. Gelnovatch, V. G., ERADCOM, "Microwave Device and Circuit Components," *Microwave Journal,* pp. 34, 36, December, 1979.

12. Cohen, E. D., NRL, "Navy Microwave Component Contracts," *Microwave Journal,* p. 35, August, 1979.

13. Cohen, E. D., NRL, "Navy MW Component Contracts," *Microwave Journal,* p. 51, February, 1980.

14. Lies, B., Raytheon, "FET's Replace Tubes in Many Design Applications," *Microwave System News,* pp. 51–60, July, 1984.

15. Tserng, H. Q., TI, "Design and Performance of Microwave Power GaAs FET Amplifiers," *Microwave Journal,* pp. 94–95, 98–100, June, 1979.

16. Ohta, K., et al, Fujitsu, "A Five Watt, 4–8 GHz GaAs FET Amplifier," *Microwave Journal,* pp. 66–67, November, 1979.

17. "Monolithic GaAs Dominates Microwaves at ISSCC," *Microwave System News,* pp. 13–32, February, 1980.

18. Fawcett, J., "Japan Charts 1980's: FET Prices to Plummet?", *Microwave System*

News, pp. 42–67, February, 1980.

19. Lain, C. M., NESC, E. J. Gersten, GE, "AN/TPS-59 Overview," IEEE 1975 International Radar Conference, April 21–23, 1975, Arlington, VA, IEEE Publicatin 75 CHO 938-1 AES, pp. 527–532.

20. Goel, J., et al, "A 1 Watt GaAs Power Amplifier for NASA 30/20 GHz Communication System," 1982 IEEE International Microwave Symposium Digest, pp. 225–227.

21. Fukuden, N., et al, Fujitsu, Japan, "A 4.5 GHz, 40 Watt GaAs FET Amplifier," 1982 MTT-S International Microwave Symposium Digest, pp. 66–68.

22. McCarter, S., et al, TI, "Design of Medium Power 6–12 GHz GaAs FET Amplifier Using High Dielectric Constant Networks," 1982 IEEE MTT-S International Microwave Symposium Digest, pp. 159–161.

23. Takagi, T., et al, Mitsubishi Electric Corp., Japan, "A 1.5 Watt, 28 GHz Band FET Amplifier," 1984 IEEE MTT-S International Microwave Symposium Digest, pp. 227–228.

24. Feng, M., et al, "Ultrahigh Frequency Operation of Ion-Implanted GaAs Metal Semi-conductor Field Effect Transistors," Applied Physics Letters, No. 44, p. 231, 15 January, 1984.

Solid State Transmit/Receive Module

for the PAVE PAWS Phased Array Radar

DONALD J. HOFT
Raytheon Company, Equipment Division
Wayland, MA

The PAVE PAWS phased array radar (AN/FPS-115) utilized for long range detection and tracking of SLBMs is presently under construction at two sites. The system utilizes a solid state transmitter and receiver. This paper will discuss the transmit-receive configuration and the transmit/receive module developed for this application.

BACKGROUND/INTRODUCTION

The PAVE PAWS phased array radar (AN/FPS-115) presently under construction at two sites in the USA, is capable of long range detection and tracking of SLBMs. **Figure 1** shows the Otis AFB, MA site during construction in early 1978. The radar(s), which are hi-rel systems meant to operate continuously, utilize 3584 active solid state modules per site (7168 total) to (1) develop the transmitted RF power with approximately 50,000 power transistors, (2) provide low-noise receiver preamplification, and (3) provide (4-bit) phase shift capability for (transmit and receive) beam steering. This paper will discuss all aspects of the module which is now in production.

SYSTEM CONFIGURATION

Figure 2 is a basic block diagram showing the transmitter and receiver-beam former (RBF) configuration for each face (2 per site). As shown, a single "array" predriver power amplifier (PA) drives 56 "sub-array" driver PAs: these (each) in turn provide the drive power to 32 output modules (one sub-array). On receive the signal, following preamplification and phase-shifting within the T/R module, is combined in a sub-array and routed through the circulator to the RBF. The array driver is redundant to enhance system reliability. The sub-array drivers and output modules are "de facto" redundant and do not require "real" redundancy. System performance is maintained with as many as 200 modules per face inoperative. Since the system transmitted signal is circularly polarized, each module output is made up of 2 signals in quadrature. **Figure 3** shows a complete T/R output module. The unit plugs directly into the array antenna element thereby eliminating the need for a coaxial line between the module and element.

Thus the insertion loss, mismatch, and most importantly the reliability are not negatively effected by this cable. The module's close proximity to the antenna is a major advantage of a solid state vs. a tube type system where

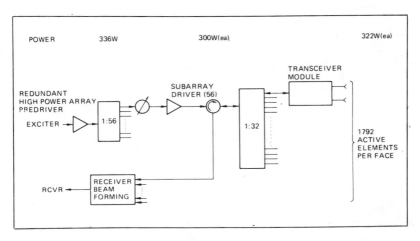

Fig. 2 PAVE PAWS Transmitter/RBF Configuration.

the losses between the amplifier and element are typically greater than 3 dB. The coaxial fittings (shown) were developed for the program. It is mechanically flexible in all three planes to eliminate tolerance buildups in the array mechanical structure.

Fig. 3 T/R Output Module

MODULE REQUIREMENT/ DESCRIPTION

Figures 4 and **5** are block diagrams of the module and power amplifier including levels, NFs, etc. **Table 1** lists the key electrical requirements. What might not be evident from **Table 1** is that the major difficulty in building solid state modules for transmitters is the high importance and difficulty of making all (and extensive) performance parameters repeat — in this case 7200 times. Major considerations in approach to the transmitter/module development were (in addition to cost):

● Conservative.(reliable) design for long life

● Manufacturability

● Maintainability: field repairable

.● Of less significance was size and weight

Figures 6 and **7** are views of the transmitter and receiver (sub) modules. To achieve these objectives the following features were implemented:

● Power transistor low junction temperatures; typically 80-120°C, 140°C max

● A 4 parallel transistor output stage in a 1-2-4 amplifier line-up

● Die cast, dual RCVR and transmitter chassis

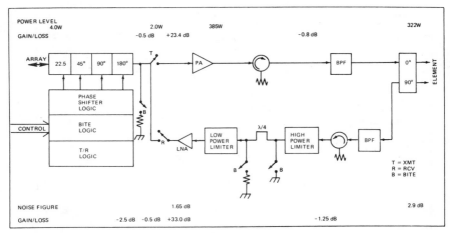

Fig. 4 SSM Block Diagram.

- Conventional PCBs/Techniques
- PCBs (7) separately testable/replaceable/repairable
- Automated manufacturing line, including RF test station
- Interchangeable dual power transistor supplies
- Semi-automated, on-site, RF test stations

With the operating junction temperatures as indicated, the MTBF (calculated) is approximately 250K and 220K hours for the receiver and transmitter (sub) modules. Each module undergoes 168 hours of RF operation after assembly and (pre) test. This is followed by final module acceptance test. Over 600 detailed measurements are made on each module in less than 7 minutes at the RF test station. One percent (1%) of all modules have their power transistors infrared (IR) scanned through clear sapphire windows (caps) to assure acceptable junction temperatures.

POWER TRANSISTORS

Two sources of power transistors are utilized (see **Figure 8**) to supply the 50,000 transistors needed for 2 sites.

Table 2 lists the major characteristics. Each transistor (source) has its own PCB design; these are interchangeable within the module. Similarly, modules are interchangeable within the radar. While the basic transistor "die" design existed, each was tailored and optimized for the PAVE PAWS application. All transistors undergo high-rel screening including 168 hr dc burn-in. Samples of every wafer lot undergo 1000 hr RF testing to qualify the wafer lot. Acceptance RF testing is done utilizing the actual module circuit; to assure tolerance to circuit and device variations, performance testing includes operation under load mismatch conditions.

PERFORMANCE

Production of the transmit/receive modules started in late 1977 and has built-up to an 800 monthly rate. Over 3 million power transistor hours have been accumulated, and results to date are excellent. **Table 3** lists electrical performance based on the first 300 modules. The results are somewhat better based on data of 3000 modules. Over 3500 modules have been produced to date.

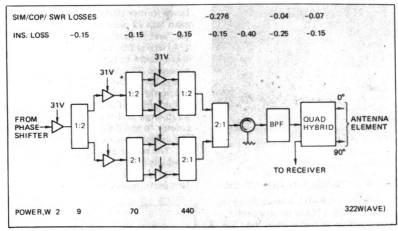

Fig. 5 Transmitter Module Configuration.

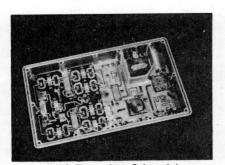

Fig. 6 Transmitter Submodule

Fig. 7 Receiver Submodule

Fig. 8 Power Transistor

ACKNOWLEDGMENT

Some of the material presented in this paper was developed under contract to the Air Force System Command's Electronic Systems Division (ESD), Hanscom Air Force Base, Bedford, Massachusetts. ⊕

TABLE 1	
TRANSCEIVER MODULE ELECTRICAL SPECIFICATIONS	
TRANSMIT	**SPECIFICATIONS LIMITS**
Frequency	420-450 MHz
RF Peak Power Output	284-440 Watts
	322 Watts Avg
Pulse Width	0.250 to 16 msec
Duty Cycle	0 to 30%
Efficiency	36% Ave (Min)
Antenna Port Tracking (Circularity)	
Amplitude	0.25 dB
Phase	3 Deg
Phase Tracking Error	14 Deg RMS
Phase Settling	25 Deg PK
Pulse Droop	1.0 dB Max
Harmonics	−90 dBc
RECEIVE	
Gain	30 dB Min ± 1 dB
Noise Figure	2.9 dB RMS Max
Limiter Power Handling	440 Watts, 16 msec, 25% D.F.
Phase Tracking	10 Deg RMS
Dynamic Range	KTBF to −28 dBm 1 dB CPRSN
No. Phase Shifter Bits	4
Phase Shifter Error	4.6 Deg RMS

TABLE 2	
PAVE PAWS POWER TRANSISTOR CHARACTERISTICS	
Sources of Supply:	Power Hybrids Inc (PHI) Communications Transistor Corp. (CTC)
Power Output (16W Pin):	110-120 Watts PK (Ave) 130 Watts PK (Max) 100 Watts PK (Min)
Duty Factor:	30% Max
Pulsewidth:	16 Millisec Max
Configuration:	PHI; Single Ended, Internally Matched In/Out CTC; Balanced, Internally Matched In/Out
Metalization	PHI; Gold CTC; Aluminum
Junction Temperature (1.6:1.0 SWR)	PHI (Au); 140°C Max CTC (Al); 120°C (Ave) Max: 10% Max May Fall Between 120-140°C
Ballast Resistors	PHI; Diffused CTC; Nichrome
Collector Efficiency	65% Ave, 60% Min
Sapphire Lids	5%
Quality	Jantxv Equivalent

TABLE 3		
MODULE PERFORMANCE (300 MODULES)		
	Performance	Spec
Power Output	330 Watts (Ave)	322 Watts (Ave)
Output Power Tracking	0.24 dB RMS	0.58 dB Max (Ave)
Insertion Phase Tracking, Transmit	6.7 Degrees RMS	14 Degrees Max RMS
Phase Shifter Error		
• Transmit	2.52 Degrees RMS	4.6 Degrees RMS Max
• Receive	2.30 Degrees RMS	4.6 Degrees RMS Max
Efficiency	37.88% Avg	36% Avg Min
Receiver Gain	34 dB Avg	27 dB Min
Receiver Gain Tracking	0.57 dB (RMS)	0.81 dB RMS Max
Insertion Phase Tracking, Receive	5.56 Degrees RMS	10 Degrees Max RMS
Noise Figure	2.71 dB (RMS)	2.9 dB RMS Max

A 250kW Solid-State AN/SPS-40 Radar Transmitter

K. Lee, C. Corson, and **G. Mols**
Westinghouse Electric Corporation
Defense and Electronics Center
Baltimore, MD

Introduction

Over the previous ten years, solid-state radar transmitter performance and cost competitiveness have been evolving to the point that now these transmitters are very common in new system and some retrofit system designs for frequencies through at least L-Band. This technology area is tied very strongly to the microwave bipolar transistor technology capability. As such, the first appearances of solid-state radar amplifiers were in preamplifier or driver stages of high power transmitters or in phased arrays where element power levels tended to be relatively small. During this time, two important trends occurred. First, the performance of these amplifiers was generally demonstrated to be highly superior to that of corresponding tube amplifiers. Bandwidth, stability and efficiency are among the documented improvements. Most importantly, however, a ten to one hundred times reliability improvement was also demonstrated.

Second, the deficiencies of both the transistor technology and the amplifier design and fabrication process technology were identified and subsequently corrected.

In the transistor area these included active area geometry, metalization and transistor packaging improvements. In the amplifier area, design architecture, circuit fabrication processes and amplifier packaging were substantially improved.

The result of these efforts was that by the late 70s designers were in a position to successfully demonstrate truly high power level solid-state transmitters. Also by this time, the decreasing cost per watt of microwave transistor power had made solid-state transmitters cost competitive to tube transmitters of equal performance.

This paper describes a 250kW peak power solid-state transmitter for use with the AN/SPS-40 naval search radar system. The development was done under contract N00014-81-C-2039 for the Naval Sea Systems Command. The solid-state transmitter design meets or exceeds all of the performance criteria of the existing tube transmitter and also provides many new capabilities. The most important of these is a 10,000 hour mean-time-between-failure. The transmitter has a single output port, as opposed to an active phased array transmitter, and operates under the existing radar pulse width and duty conditions, as a minimum. A key feature of the transmitter is that a high degree of modularity was used: 128 total amplifier modules comprised of only two types, and 12 identical power supplies.

Transmitter Architecture

The nominal 250kW output of the transmitter is achieved by the parallel combination of 112 transmitter modules located in the output stage of the transmitter. Each of these amplifier modules has a nominal output power of 2500 watts. As seen in the transmitter block diagram shown in Figure 1, the 112 output stage modules are arranged into two groups of 56 each. Each of these groups is driven by a single input power level of seven kilowatts and is combined separately to 130kW level. The two groups are then sation for temperature, frequency combined together to produce 250KW. The reason for this grouping is that each is physically housed in its own power amplifier cabinet, as shown in Figure 2. These two output cabinets are driven by a three stage driver chain which is completely housed in a third cabinet, the driver cabinet.

The drive stage and the pre-driver stage use twelve and two 2500 watt modules respectively, each of these being identical to and interchangeable with the 112 output modules. The first stage, the preamplifier, uses two parallel 500 watt high gain modules. The modues are identical to each other but are different from the 2500

watt modules. Thus, one of the key transmitter architecture features can be seen. Namely, the total count of unique types of modules can be minimized by the correct selection of intermediate stage power levels. This is often desirable from at least maintainability and system logistics standpoints. Each drive stage also has its appropriate splitter and combiner pair. The outputs of these three stages are 17.5kW, 1700 watts and 250 watts respectively.

A comparison between these power levels and the potential power levels of each stage (the algebraic sum of the intra-stage module capabilities) points to another key architecture characteristic. This is, that each stage has the potential to generate much higher power than that required, and therefore when properly controlled can maintain the desired stage output level in the presence of partial failures within that stage. Also, compenand module to module performance variation can be achieved. Certainly this seems desirable in general but is especially important in solid-state radar transmitters.

These transmitters predominantly use bipolar, class C transistor circuits and use many of them in parallel to achieve the desired output power. In a large transmitter, several series stages are also needed. If the proper drive level into each stage is not kept close to its design value, 0 to -2 dB. the stage will exhibit a gain loss which will be magnified by subsequent stages in turn. The result is that a relatively small

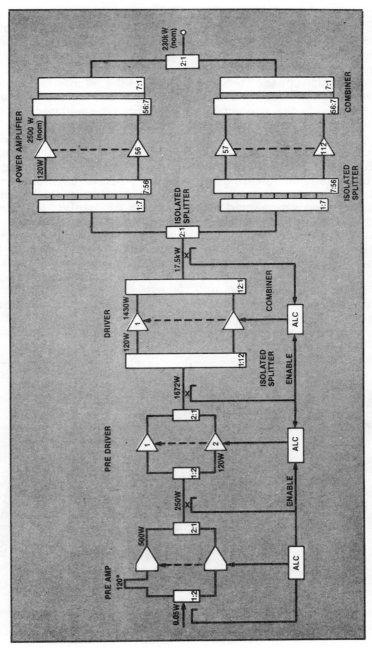

Fig. 1 Transmitter block diagram.

degradation, if uncorrected, in the drive stages may cause large amplitude variation in the output. In the cases where subsequent stages use parallel circuits, instabilities may also appear. Thus, the Automatic Loop Control (ALC) circuits provide the needed adjustment to overcome these potential problems. These circuits control the bias voltage to the individual stage amplifiers which in turn control the microwave gain.

As shown in Figure 1, each of the drive stages is monitored at its input and output for RF amplitude. These signals are processed in a low noise digital fashion and the output error signal is used to control the stage bias voltage. In addition to these techniques, 100 percent amplifier redundancy is provided in the preamplifier stage, where two 500 watt amplifiers are combined in a 120° phase relationship. Here, either amplifier alone or the two working together will produce the desired 250 watt output power.

All of the microwave amplifier modules are biased by twelve identical programmable 24 to 40 volt DC power supplies. This power system is modularized in a manner similar to the microwave amplifiers. Each of the output power amplifier cabinets has five supplies, each powering either eleven or twelve 2500 watt modules. In this approach, if a supply fails the output power will gracefully degrade by only 1 dB until it can be replaced. Two additional supplies are used to power the driver stages, one on line and one 100 percent redundant on active standby. Either one can power the entire driver chain alone.

Fig. 2 250 kW solid-state transmitter.

Cooling of the transmitter is accomplished in the normal mode by a combination of forced air and liquid coolant. The power supplies and auxiliary circuitry are cooled by air at all times. The modules are cooled by conduction to a cold plate which in turn is normally liquid cooled. A backup air cooled mode is also provided for the cold plate in case of a liquid coolant system failure. All cooling systems are redundant, and therefore eliminate a major source of general transmitter non-availability.

The electronic brain and human interface for the transmitter are housed in the driver cabinet. Here the inputs from the transmitter's 160 sensors in the RF amplifier, power supply and cooling subsystem are processed and displayed. Also, self-healing is automatically controlled from this unit with provisions for manual override provided. Figure 3 is a picture of the transmitter's control/display panel. The data shown here is also summarized for transmission to remote system control electronics.

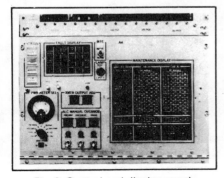

Fig. 3 Control and display panel.

Transmitter Power Modules

The design of the transmitter power modules is a critical task in all solid-state transmitters. It forms the basis around which the rest of the overall design will center. At issue in the basic module approach are such considerations as module complexity, power level, total transmitter module count and cost, graceful degradation and maintainability. Generally, overall system complexity and cost are lower if fewer modules are used. In the extreme, overly large and complex modules may defeat these advantages and introduce less desirable levels of graceful degradation and maintainability to the transmitter. As an example, if one of the 112 output modules in the AN/SPS-40 solid-state transmitter should completely fail, the total output power would decrease by only 0.08 dB. If the transmitter were to have 1/10 the quantity, but proportionately higher power modules and if one should fail, the output power would degrade by 0.08 dB. This may not be acceptable to the user and realizing these higher power modules would be very difficult.

The module design is also a critical task since until its basic performance is verified, the entire transmitter concept has risk. Such parameters as gain, output power, stability, transistor junction temperature and inter-module performance repeatability must be verified to be within acceptable bounds. These and other parameters were considered and verified for the power amplifier module which was used in the transmitter. The SPS-40 power amplifier module is shown in Figure 4.

The module consists of two stages cascaded to amplify an input power of 120 watts peak to an output power of 2500 watts peak. There are a total of ten transistor circuits in the module. The module's block diagram is shown in Figure 5. The 120 watt peak input pulse is divided into two 60 watt pulses by a 1:2 quadrature splitter. Each 60 watt pulse is amplified by a single stage transistor circuit and the two outputs then re-combined by a complementary 2:1 combiner to produce a peak power of approximately 620 watts. The output of the combined driving transistors is then split eight ways to drive the eight output transistors. The eight outputs are again in quadrature. This relationship between adjacent paralleled transistors is important since it tends to isolate preceding circuitry from the voltage standing waves normally created at the inputs of the transistors. In doing so, overall stability and uniform frequency performance are maintained. The eight splitter output signals are then amplified by eight transistors and recombined by an 8:1 complementary combiner. The resultant power level is nominally 2700 watts.

A sampling coupler is located after this point and its output is used to establish a performance monitor signal for the module. This signal is provided for auto-

Fig. 4 2500 W power amplifier module.

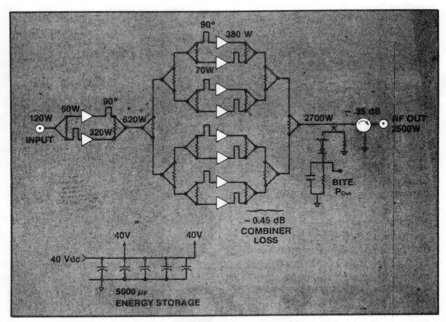

Fig. 5 Power amplifier module block diagram.

matic performance monitoring and fault isolation at the transmitter level. Following the sampling coupler is an isolator which protects the rest of the module circuitry from voltage standing waves which may be generated external to the module.

The power amplifier module is a hermetically sealed, plug-in module. A single sided circuit board (aluminum class soft-substrate E = 10.2) provides most of the module's microstrip circuitry. A capacitor bank, circulator, and various transition pieces make up the rest of the subassemblies for the PA module. Two quick disconnect hermetic connectors are used for the input and output of

RF to the module. A multipin DC connector along with hermetic EMI filters serve as the electrical paths for DC power to the module and the performance monitoring output signal. Shear pins are located at the rear of the module to provide support for mechanical shock the module may encounter during operation. The shear pins also help guide the module into its appropriate position in the transmitter cabinet and two captive fasteners are used to secure the module.

The basic building block for the module is a 400 watt peak power transistor of which there are ten in each module. The primary characteristics of this transistor are listed in Table 1.

One of the major goals of the design was to evolve a power module design which could successfully use more than one source of RF power transistors. The benefits of this are: guaranteed production quantity flow for original module manufacture, a low unit production price due to multiple competitors and a secure supply of long term spare parts. To attain these benefits, a direct one-for-one substitution of parts manufactured by different vendors was desired. At the beginning of the design two vendors, Communication Transistor Corp. (CTC) and Power Hybrids Inc. (PHI), had transistors which basically met the performance requirements. However, they were not at all interchangeable, since one was a common base and the other a common emitter part. Working closely with these vendors and a third, Thompson CSF (T-CSF), a common specification and circuit was established which would serve as the baseline for all transistors to be used in the transmitter. After

several iterations of circuit and transistor designs, a state-of-the-art level of interchangeability was accomplished while still maintaining exceptional performance. The end results can be seen graphically in Figure 6. Represented here is the output power performance versus frequency of a typical module. Curves A, B, and C fall on top of one another and represent performance under the following conditions:

A - All module transistors from one vendor
B - 9 transistors from one vendor and 1 from a second
C - 8 transistors from one vendor and 2 from a second.

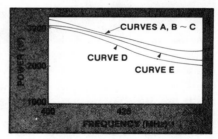

Fig. 6 Transistor interchangeability.

TABLE I	
TRANSISTOR KEY CHARACTERISTICS	
Frequency	400 MHz-450 MHz
Output Power (Peak)	400W-500W
Efficiency	55% Min.
Gain	8 dB Min.
Metalization	Gold
Pulse Width	60 µS
Duty	5%
Package	Hermetic

These substitutes were made by replacing only the transistor(s) with no subsequent retuning of the circuitry. Curves D and E show slight performance degradation but still within specification under the following conditions:

D - 9 transistors from one vendor and 1 from the third vendor

E - 7 transistors from one vendor, 2 from a second vendor, and one from a third vendor.

Other tests on the modules also showed that as the transistors were substituted, all other module performance criteria were also met. Among these are insertion phase length, efficiency and the junction temperature of the transistors themselves. The nominal junction temperature for the transistors is 95°C peak and the worst case temperature is 125°C. A photo of all three transistors used in the transmitter is shown in Figure 7. The circuits are shown in the schematic, Figure 7a.

There were 128 of the power amplifier modules fabricated for the transmitter. Almost all of these have a homogeneous selection of transistors within them. Modules were made with each vendor's transistor, but for this first transmitter, mixing within modules was kept to a minimum. The reason for this was that this approach would make transistor reliability data less ambiguous if a particular vendor's device had a higher failure rate during transmitter evaluation and life testing. Uniform performance of the 128 modules is

shown in Figure 8. Another indication of uniformity among the modules was indicated by subjecting a random sample of modules made with transistors supplied by various vendors to temperature testing. These experiments showed a worst case amplitude tracking of only 0.2 dB and phase tracking of only ±5° over an 80°C ambient temperature range.

Power Combiners

Another key to very high power solid-state transmitters is obviously the power splitting and combining of the signal entering and leaving the modules. While the theoretical aspects of these circuits have been solved for some time, the practical problems are still being worked today. The two largest considerations which make the combiners a particularly difficult area are:

1. providing the absolute lowest loss possible and,
2. providing the power handling capability.

For some time the solution to these concerns has been to use an air dielectric medium to realize the circuitry.[1, 2] This has the dual advantage of lowering the dielectric loading loss found with solid dielectrics and increasing the voltage breakdown capability, particularly at transition points. This is so since the unity dielectric constant increases circuit spacings relative to the solid dielectrics. Several technologies have been used successfully to accom-

Fig. 7 SPS-40 power transistors.

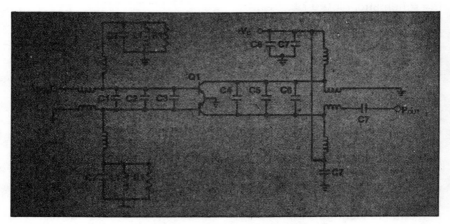

Fig. 7a Transistor circuit schematic.

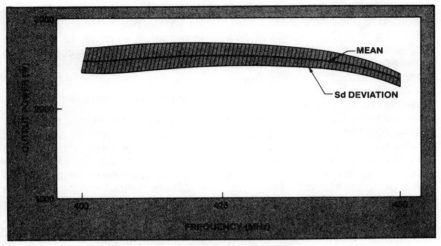

Fig. 8 Module performance.

plish this approach: stripline, cavity, coax and waveguide.

The large output 56:1 combiners for the AN/SPS-40 solid-state transmitter use the stripline approach. This was chosen since it has a very good form factor and is extremely predictable in design. The dimensions of the 56:1 combiner are 60" x 19" x 2" and its component pieces are entirely made by automatic computer controlled machines to provide low cost and reproducibility. A picture of one of the combiners is shown in Figure 9. The 56:1 combiner is a non-isolated device which contains two layers of stripline. The input layer consists of seven 8:1 impedance-type combiners mounted in a 0.5 inch ground plane spacing. These are combined in a 7:1 combiner which has a ground plane spacing of 1 inch. The wider spacing is necessary

due to the power levels involved after the 8:1 combinations. This layer's output is a 3-1/8 inch diameter coax transition which has a standard EIA-type flange. The nominal output power of the combiner is 130kW and its maximum loss is an impressive 0.25 dB as shown in Figure 10.

The combiner was designed with a large degree of operating safety margin, which was the key to success of the transmitter's development. The final design was successfully operated at a 250kW power level, which is twice its intended usage level, into SWRs ranging up to infinity. Also important is that the combiner was operated as a combiner. The usual way of testing a combiner is to drive their single output port and measure the input ports one by one to establish loss and phase match data. For the high power test of the 56:1 however, it was decided to use another 56:1 combiner as a splitter, join the two together and run composite tests. This established the true dynamic loss due to both amplitude and phase. It also provided a means to test the combiner under abnormal but real life conditions. One of these conditions is the ability to withstand the incident power levels when a various number of the input ports are either not driven and/or mismatched. The final design passed all these tests and under normal operating conditions has a 4:1 power safety margin when operating into open or short circuit loads.

The output power of the two 56:1 combiners are combined in a 2:1 3-1/8 inch coaxial combiner which is mounted external to the cabinets. The loss of this combiner is less than 0.1 dB.

Transmitter Capabilities

The key specifications of the solid-state transmitter are shown in Table 2. Among these is the non-typical capability of continuous output power availability in 10 kW increments from 0 to 250 kW. This feature is controlled by switches provided on the control/display panel shown in Figure 3. As the self-healing driver chain is controlled, as described above, the output power can also be con-

Fig. 9 56:1 combiner.

trolled by varying the command to the power amplifier cabinets' power supplies.

Another capability is that of a backup air cooling mode for the modules should the ship's liquid cooling system fail. As seen in Figure 11, the transmitter's output power in the total air cooled mode is slightly lower but still in excess of 200 kW. This feature is very important for availability and would be used in emergency situations when the extra thermal load in the transmitter compartment would be acceptable to the user.

Besides being extremely modular, almost all of the transmitter subassemblies are quick plug-in units. All of the RF modules, module power supplies, splitters, combiners and PC boards are plug-in, front-access assemblies

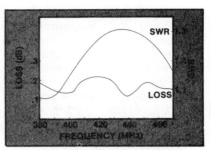

Fig. 10 56:1 combiner parameters.

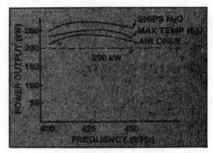

Fig. 11 Output power vs. frequency.

TABLE II	
KEY TRANSMITTER SPECIFICATION	
Frequency	400 MHz to 450 MHz
Instantenous Bandwidth	50 MHz
Output Power Peak	200kW to 250kW nominal Adjustable from zero to full power
Output Power Average	4kW. nominal
Bandpass Flatness	±.5 dB maximum
Gain	66 dB minimum
RF Load SWR	1.5: nominal No damage at ∞:1
Primary Power	440V RMS, 3ϕ, 60 Hz 115V RMS, 1ϕ, 60 Hz
Cooling	Forced Air and Liquid

with all connections made automatically with the exception of RF connections. These, if appropriate to the assembly, are generally made at the front of the assembly. In addition, the modules and power supplies can be replaced while the rest of the transmitter is operating.

Thus, 75 percent of all predicted transmitter maintenance items can be accomplished in a true on-line fashion.

The transmitter output power sensitivity (graceful degradation) to various failures is shown in Table 3. The excellent numbers are due to the degree of modularity, redundancy and self-healing included in the design. Of course, the transmitter also has the normal features of a solid-state design in that it is an instant-on unit with no warm-up time required and has the full operational bandwidth in an instantaneous sense.

Conclusions

Very high power solid-state transmitters are certainly feasible and cost-effective for radar systems. Properly designed and tested they provide at least equal performance and far superior reliability, availability and maintainability compared to their tube counterparts. By utilizing their inherent modularity, redundancy and self-healing can be provided in a fashion which is very cost effective and straight-forward.

The key subsystem components are and will continue to be the amplifier modules and the high power combiners. However, substantial effort must also be placed into the areas of overall architecture, power supplies, BITE, automatic control and cooling for a given transmitter design to be successful. As we continue to evolve into bigger and better solid-state transmitters, overall system and transmitter architecture will play a large role in the acceptability of these efforts.

Acknowledgments

The authors would like to thank Mike Lipka of the Naval Research Laboratory for his assistance and support during this development. Also to be thanked are the large team of people at both Westinghouse and various Navy agencies for their efforts in support of this program.

TABLE III	
TRANSMITTER GRACEFUL DEGRADATION AND SELF-HEALING	
Failed Unit	**Output Power Loss**
Driver Module	None
Preamplifier Module	None
Driver Power Supply	None
Output Module	0.08 dB
Output Power Supply	0.9 dB
Fan Failure	None
Liquid Failure	0.5 dB

Appendix A

Proof of the CSA Properties

1. A minimum number of unit amplifiers is needed to obtain a given output power when the fan out ratio is equal to the unit amplifier power gain.

Proof:

The power output $P_o = L_c G^K P_u$ and G^K output stages are used. With fan-out ratio G, the number of driver stages is

$$N_d = \frac{G^K - 1}{G - 1}$$

If the same power output is produced with a fan-out ratio $(G-1)$, the number of driver stages is

$$N_d' = \frac{(G-1)^P - 1}{G - 2} = \frac{G^K - 1}{G - 2}$$

since $(G-1)^P = G^K$ (the number of output stages is constant), therefore,

$$\frac{N_d'}{N_d} = \frac{G - 1}{G - 2}$$

which is greater than 1 for all values of G. Because the number of drivers is greater and the number of output stages is the same with fan-out ratio $(G-1)$, fan-out ratio G uses fewer devices.

With fan-out ratio $(G+1)$, the drive power developed by the first stage is divided by $(G+1)$ and, since its power gain is G, the drive to each stage in the next rank is $[G/G+1]P_{in}$. The drive power at rank K is $[G/G+1]^K P_{in}$. The drive power required is at least P_{in}. This will be satisfied with a fan-out ratio less than or equal to G.

2. For transistor types with the same gain and different power output ratings, a CSA utilizing the highest output power devices from the above set will require a minimum number of transistors to produce a given power output.

266

Proof:

Consider a group of power transistors with the same gain, but with output powers P_j that are different. Because all have the same gain, a CSA incorporating these devices will have the same overhead ratio

$$O = \frac{1-G^K}{G-1}$$

For a required output power P_o, the number of devices in the output rank P_o/P_j. This number is smallest with the largest P_j. Because the overhead ratio is the same for all cases, the CSA that uses the devices with the highest power output will employ the lowest number of devices to provide the required output.

3. When a given power output is developed by summing the output of two or more CSAs the number of transistors used is the same as would be required in a single CSA with the same power output and power gain.

Proof:

A CSA of rank K produces a power output P_o. Each unit amplifier produces P_u watts and has a power gain G. The power gain of the CSA is G^{K+1}. If this CSA is replaced by a quantity G of rank $(K-1)$ CSAs, the output power will be the same. The number of devices in the rank K CSA from (4-2) is

$$N_t = \frac{G^{(K+1)} - 1}{G-1}$$

The number of transistors in the G CSAs of rank $K-1$ is

$$N_t' = G[G^K - 1]/[G-1] \quad \text{therefore } N_t - N_t' = 1$$

but the gain of the G CSAs in parallel is G^K. To make the group equivalent to the rank K CSA, one additional stage at the input will be needed. Therefore, the total number of unit amplifiers is identical.

4. When a CSA is built using modules which contain more than one unit amplifier, the fan-out ratio between modules that is equal to the module power gain will produce the required output power with a minimum number of modules. If the module is also a CSA, the number of transistors required to produce a given output power for the modular CSA is identical to the number used in a CSA composed of unit amplifiers that produces the same output power.

Proof:

A CSA transmitter of rank K is composed of modules with power output P_m and power gain G_m. Each module is also a CSA of rank r composed of unit amplifiers with power gain G and output power P_u. The output power of the transmitter is

$$P_o = L_c G_m^K P_m = L_c' G_m^K G^r P_u$$

but

$$G_m = G^{(r+1)}$$

and

$$P_o = L_c' G^{(kr+k+r)} P_u$$

This is the same power that would be produced by a unit amplifier based on a CSA of rank $[(r+1)(k+1)-1]$. The number of unit amplifiers in each module is

$$N_u = \frac{G^{(r+1)} - 1}{G - 1}$$

The number of modules in the transmitter is

$$N_m = \frac{G_m^{(K+1)} - 1}{G_m - 1}$$

Substituting $G^{(r+1)}$ for G_m, we have

$$N_m = \frac{G^{(r+1)(k+1)} - 1}{G^{(r+1)} - 1}$$

The total number of unit amplifiers in the transmitter is

$$N_t = \frac{G^{(r+1)(k+1)} - 1}{G - 1}$$

Therefore, the number of unit amplifiers in the modular CSA is equivalent to a unit amplifier based CSA of rank $[(r+1)(k+1)-1]$.

INDEX

ABCD Matrix, 164
AN/FPS-115, 136
AN/SPS-40 , 215, 251-264
AN/TPS-59 , 215-229
Alumina, 67-69
Amplifier stability, 108
Amplifier, balanced, 109, 111
Acquisition Cost, 209
Average Power, 155, 160-162

Balanced Amplifier, 109
Ballast techniques: 12,13
Base resistance, 61
Beryllia,(Be O), 67-69
Bias Components,106-107
Bipolar microwave transistors, 7,46,47,51
Bond failures, 30
Breakdown enhancement, 13
Built In Test Equipment,(BITE), 183

Cellular construction,13
Ceramics, 67, 68
Circulator,Isolator,148
Class C operation, 52, 53, 55
Coaxial Line Impedance, 158
Coaxial line, 157
Cohen, 151, 160
Collector efficiency, 24, 53
Collector load contour, 116-123
Collector-base capacitance , Cob, 61
Combiner isolation, 180
Corporate structure amplifier (CSA), 131
Corporate structure, 131
Cost Trade-offs, 140,206-214

DC Gain (h_{FE}), 41-42
Device geometries,9
Droop, Pulse, 196
Duty Cycle, 15, 27
Duty factor,15,27

Effective emitter area, 23
Effective resistance, 67
Efficiency, transmitter, 135
Emitter Ballast, 5, 24
Equivalent Series Resistance (ESR) Of Capacitor, 198
Field Effect Transistor (FET), 47-48

Fortran, 169

Ga As Fet, 34,47, 51
Gallium arsenide, 69,239
Getsinger, 160
Gold wire, 67
Gysel divider, 172, 177
Gysel, 151,172,177

Hermetic sealing, 22
High Temperature Reverse Bias,(HTRB), 44
High current BVces testing, 17,23, 45
Howe, 16
Hybrid Combiner, 144

Impedance matching ,internal,6
Infrared scanning microscope, 111
Insertion Loss, 158, 161-163, 165
Insertion phase, 18,165
Interdigitated geometry, 10
Internal inductance, Lint, 63
Isolation loss diagram, combiner, 180

Junction Temperature, 19, 109, 111, 124

Life Cycle Cost, 212
Load Mismatch Effects, 124
Lumped Element Design, 78-81

Maintainability, 220
Maintenance Cost, 210
Maintenance, 203
Matrix geometry,11
Mean time between failures, (MTBF), 203
Metalization failures, 29
Metalization peeling, 29
Microstrip, 87,101,160
Microwave monolithic integrated circuit (MMIC), 239-240
Modules, 183, 233, 235, 238
Moving Target Indicator (MTI), 193
MOS FET , 34, 47
MOS capacitors,20, 31
MTI Improvement Factor, 198

Operating Cost, 209
Overhead Ratio, 134
Overlay geometry,9

PAVE PAWS ,136-138,215,234,246-250
Package failures, 31
Package parasitics,6
Parad, 150
Peak Power,155-160, 161, 163
Phase Pulling and Pushing, 189-190
Phase settling, 18
Phased array, 208
Planar Divider, 172, 174, 175, 178-181
Power Adder, 143
Power Amplifier Module, 183
Power Combiner, 143
Power Gain, 17
Power Output Control, 141-142

Quadrature Hybrid, 188

Reactive Combiner, 144
Reflection Angle, 165
Reliability, transmitter, 201-207, 228
Rozenzweig, 160

S-parameters, 53
Sanders, 151
Sapphire, 69
Semiconductor chip, 7
Silicon,
Smith Chart, 84-86
Static Induction Transistors ,SITs),48
Strip line, 159

Terminating Resistor, 147-148
Thermal Radiation Equation, 159
Thermal conductivity ,69
Thermal resistance, 114
Thermal resistance, 25
Thermal transients, 27, 115
Thump, 193
Titanium Tungsten, (Ti W), 7
Transistor :
        BVceo , 40
        BVcer , 40
        BVces testing, 17,23,45
        RF modules, 20
        bandwidth, 18
        dc burn-in , 44
        dc characteristics,16
        dc gain, 41
        dice, 7
        doping,7

Transistor (cont'd):

        efficiency,18, 42, 135, 136
        failure mechanisms, 28
        failure modes,6
        insertion phase, 18
        junction temperature, 19
        leakage currents, 17, 41
        package parasitics, 19
        packaging, 19
        power gain ,17
        pulse droop, 18
        reliability , 28
        stability, 18, 36
        thermal properties, 21
Transmission line Parameters, 146
Transmission line matching, 100
Transmission lines, 94
Transmitter efficiency, 135
Travelling wave tube (TWT), 73

Uniform Corporate Structure Amplifier, (UCSA), 132

V MOS ,34, 47
Voltage Gradient, 156
Voltage Reflection Coefficient, 165
Voltage Standing Wave Ratio (VSWR), 165
Voltage Transmission Coefficient, 164

Waveguide, Double Ridged, 163
Waveguide, Rectangular, 161
White, 160
Wilkinson combiner, 149,150
Wire Pull Tests, 30

Y-Matrix, 165

Z-Matrix, 167